Risk assessments questions and answers: a practical approach

Pat Perry

Published by Thomas Telford Publishing, Thomas Telford Ltd, 1 Heron Quay, London E14 4JD.
URL: http://www.thomastelford.com

Distributors for Thomas Telford books are
USA: ASCE Press, 1801 Alexander Bell Drive, Reston, VA 20191-4400, USA
Japan: Maruzen Co. Ltd, Book Department, 3–10 Nihonbashi 2-chome, Chuo-ku, Tokyo 103
Australia: DA Books and Journals, 648 Whitehorse Road, Mitcham 3132, Victoria

First published 2003

Also in this series from Thomas Telford Books
Construction safety: questions and answers. Pat Perry. ISBN 0 7277 3233 1
Health and safety: questions and answers. Pat Perry. ISBN 07277 3240 4
Fire safety: questions and answers. Pat Perry. ISBN 07277 3239 0
CDM questions and answers: a practical approach 2nd edition. Pat Perry. ISBN 0 7277 3107 6

A catalogue record for this book is available from the British Library

ISBN: 0 7277 3238 2

Any safety sign or symbol used in this book is for illustrative purposes only and does not necessarily imply that the sign or symbol used meets any legal requirements or good practice guides. Before producing any sign or symbol, the reader is recommended to check with the relevant British Standard or the Health and Safety (Safety Signs and Signals) Regulations 1996.

Throughout the book the personal pronouns 'he', 'his', etc. are used when referring to 'the Client', 'the Designer', 'the Planning Supervisor', etc., for reasons of readability. Clearly, it is quite possible these hypothetical characters may be female in 'real-life' situations, so readers should consider these pronouns to be grammatically neuter in gender, rather than masculine.

Typeset by Alex Lazarou, Surbiton, Surrey
Printed and bound in Great Britain by MPG Books, Bodmin, Cornwall

Biography

Pat Perry, MCIEH, MIOSH, FRSH, MIIRM, qualified as an Environmental Health Officer in 1978 and spent the first years of her career in local government enforcing environmental health laws, in particular health and safety law, which became her passion. She has extensive knowledge of her subject and has served on various working parties on both health and safety and food safety. Pat contributes regularly to professional journals, e.g. *Facilities Business*, and has been commissioned by Thomas Telford Publishing to write a series of health and safety books.

After a period in the private sector, Pat set up her own environmental health consultancy, Perry Scott Nash Associates Ltd, in the latter part of 1988, and fulfilled her vision of a 'one-stop shop' for the provision of consultancy services to the commercial and retail sectors.

The consultancy has grown considerably over the years and provides consultancy advice to a wide range of clients in a variety of market sectors. Leisure and retail have become the consultancy's major expertise and the role of planning supervisor and environmental health consultant is provided on projects ranging from a few hundred thousand pounds to many millions, e.g. new public house developments and major department store refits and refurbishments.

Perry Scott Nash Associates Ltd have strong links to the enforcing agencies; consultants having come mostly from similar backgrounds and approach projects and all the issues and concerns associated with legal compliance with pragmatism and commercial understanding.

Should you wish to contact Pat Perry about any issue in this book, or to enquire further about the consultancy services offered by Perry Scott Nash Associates Ltd, please contact us direct at:

Perry Scott Nash Associates Ltd
Perry Scott Nash House
Primett Road
Stevenage
Herts
SG1 3EE

Alternatively phone, fax or email on:

Tel: 01438 745771
Fax: 01438 745772
Email: p.perry@perryscottnash.co.uk

We would also recommend that you visit our website at:
www.perryscottnash.co.uk

Acknowledgements

My sincere thanks go to Maureen for her never ending support and encouragement and to Janine and the Business Support team at Perry Scott Nash Associates Ltd for typing all the handwritten manuscripts with such patience and efficiency.

Author's note

Many of the publications referenced in this book are available for download on a number of websites, e.g.:

- www.hse.gov.uk/pubns/index.htm
- www.hsebooks.co.uk

Also, guidance on the availability of books is available from the HSE Info Line on 0541 545 500.

Contents

1

Introduction

What does the term 'risk assessment' mean?

Quite literally, it means an assessment of the likelihood of something going wrong or affecting somebody or something in a way which could cause them harm, or damage property, etc.

Risk assessment is a logical approach to reviewing the dangers in a job, activity or event and determining the consequences. The consequences may be acceptable to the Assessor or they may not. If the dangers of a job or activity are considered to be too great, then the activity should not be undertaken unless control measures are put in place, i.e. steps are taken to reduce the risk of harm or injury to an acceptable level.

Why is risk assessment so important in respect of health and safety?

Employees and others undertake a vast array of jobs and activities in their working environment and it is important that employers protect their safety at all times. In order to know what the hazards and risks are in respect of a specific job activity, the employer will need to assess the job so as to be able to identify them.

Employers need a systematic method of reviewing and recording the identified hazards from a job and the risk assessment process provides this.

Is risk assessment something new?

No. There has been an implied requirement to carry out risk assessment in the Health and Safety at Work Etc. Act 1974, as Section 2 requires employers to take steps to ensure the safety of employees and others 'as far as is reasonably practicable'. In order to assess what is 'reasonably practicable' it has always been necessary to review practices, procedures, tasks and jobs to find out what is being done and what can be done to ensure the safety of individuals.

The first formal requirement for risk assessment under legislation was probably in the Control of Substances Hazardous to Health Regulations issued in 1988. The Regulations stated that an employer had to make a 'suitable and sufficient assessment of the risks created by work liable to expose any employee to any substance hazardous to health'.

Less prescriptive requirements for risk assessment were contained in the Control of Lead at Work Regulations 1980 and the Control of Asbestos at Work Regulations 1987.

Is risk assessment as complicated as everyone seems to believe?

No. Risk assessment is a common sense approach to identifying the hazards and risks associated with a work task or activity.

Some knowledge of the task or job process is needed as is some understanding of what could go wrong, what harm could be caused and to whom and how likely it is to happen.

Risk assessments need to be carried out by 'competent people' not fully trained, degree level experts.

Some aspects of risk assessment are complicated but these are usually where the work environment or processes undertaken are complex and potentially very dangerous, e.g. oil rigs, railways, chemical processing plants.

What is meant by the term 'suitable and sufficient'?

Exactly what the words mean — suitable for the level and complexity of the job and sufficient in that it identifies as many of the known hazards and risks as possible.

A risk assessment needs to reflect the key elements of the job or activity and must consider all known consequences.

People are not expected to be able to 'see round corners' and sometimes the level of knowledge of a particular hazard or its potential has not developed, for example no one really knows at the moment whether the use of mobile phones poses a real health hazard but perhaps, in a year or two, research may conclude that their use is a considerable health hazard. Employers are not expected to implement controls to hazards that are not fully understood.

Suitable and sufficient is based on the knowledge we have at the time of undertaking the risk assessment.

What does the term 'hazard' mean?

Hazard is the 'potential to cause harm'.

Virtually anything has the potential to cause harm if used inappropriately or in an unsuitable environment. Life is full of hazards — all manner of things *could* cause us harm, e.g. using electricity, crossing the road, driving the car, walking, playing sport, etc.

Many work activities have the potential to cause harm to employees and others.

However, just because a hazard exists does not mean that any harm will necessarily result because certain activities are undertaken to reduce the consequences of hazard.

What is meant by the term 'risk'?

Risk is the likelihood that the hazard will be realised or the harm identified from the hazard will come to fruition.

Risk is what might happen — the consequences of the harm or hazard.

Risk can affect one person, several people or hundreds and the number potentially involved helps to determine the seriousness of the risk.

What is meant by the term 'risk control'?

Risks are generally unacceptable — why accept the potential to be injured or harmed if you do not have to. Risk control is about managing the risks so that they become acceptable.

Risk control is the outcome of a risk management process.

What is meant by the term 'risk reduction'?

Risk reduction is a management tool to be used to reduce the likelihood and consequences of risks throughout an organisation.

Records, statistics, insurance losses, etc. can be reviewed to see what types of hazards and risks are happening and the costs to the business. A review will then take place and an objective will be set to reduce the risks inherent in particular activities, processes, etc. so that the losses incurred will reduce.

Risk reduction reduces the likelihood of things going wrong and reduces the consequences and costs from those risks.

What is risk avoidance?

Usually, risk avoidance is a conscious decision on the part of the employer (or other) to avoid any risks associated with a job or activity.

Why accept the potential for harm if you do not have to? A risk avoidance strategy identifies hazards and risks and states that those risks are unacceptable and will not be tolerated. Some alternative way of completing the job or activity without incurring risk will be found.

Working at heights is a hazard and there is a risk of falling from the height, dropping objects or tools onto people below, etc. An employer may adopt a 'risk avoidance' strategy by ensuring that all work at height is from a level platform with no risk of falling, e.g. windows cleaned from the inside because they have been designed with 360° centre pivots.

What regulations in respect of health and safety require risk assessments?

The concept of risk assessment has developed over the years and has become an important part of UK health and safety legislation.

New health and safety laws are generally known as 'self-regulatory' as opposed to being 'prescriptive'. This means that the Regulations no longer tell employers what they must do and how, but set a framework for achieving 'goals' in respect of health and safety so that the employers can arrive at a solution which suits their businesses. The risk assessment approach enables this to happen by, for instance, stating that employees will not be exposed to risk but allowing the employer to determine what that risk is and how to reduce it, etc.

The following Regulations require risk assessments specifically:

- Noise at Work Regulations 1989
- Manual Handling Operations Regulations 1992

- Health and Safety (First Aid) Regulations 1981
- Health and Safety (Display Screen Equipment) Regulations 1992
- Personal Protective Equipment Regulations 1992
- Fire Precautions (Workplace) Regulations 1997 and 1999
- Control of Substances Hazardous to Health Regulations 2002
- Control of Asbestos at Work Regulations 2002
- Control of Lead at Work Regulations 2002
- Dangerous Substances and Explosive Atmospheres Regulations 2002
- Ionising Radiations Regulations 1999
- Control of Major Accident Hazard Regulations 1999.

And, of course, the Management of Health and Safety at Work Regulations 1999 which set the framework and expectation for all risk assessments. If risk assessments are not called for under specific Regulations the requirement to complete them in any event is contained in the Management Regulations.

What are the consequences for failing to carry out risk assessments?

The ultimate consequence for failing to carry out a risk assessment is that an employee or other person could be killed by the job or activity you have asked them to carry out as their employer.

Accidents and injuries to employees or the general public caused by work activities can lead to compensation claims through the civil courts or insurance companies.

Prosecutions for failing to have suitable and sufficient risk assessments are common and, depending on the circumstances of the case, the case can be heard in the magistrates' court or Crown Court.

Fines can be up to £5000 for each missing or inadequate risk assessment or they can be unlimited in the Crown Court if, for instance, the lack of a risk assessment contributes to a fatality.

2

Legal framework

What is the main piece of legislation which sets the framework for health and safety at work?

The Health and Safety at Work Etc. Act 1974 is the main piece of legislation which sets out the broad principles of health and safety responsibilities for employers, the self-employed, employees and other persons.

The Health and Safety at Work Etc. Act 1974 is known as an 'enabling act' as it allows subsidiary Regulations to be made under its general enabling powers. It sets out 'goal objectives' and was one of the first pieces of legislation to introduce an element of self-regulation.

The Act places responsibilities on employers to:

- safeguard the health, safety and welfare of employees
- provide a safe place of work
- provide safe equipment
- provide safe means of egress and access
- provide training for employees
- provide information and instruction
- provide a written Safety Policy
- provide safe systems of work.

The Act places responsibilities on employees to:

- co-operate with their employer in respect of health and safety matters
- wear protective equipment or clothing if required
- safeguard their own and others' health and safety
- not recklessly or intentionally interfere with or misuse anything provided in the interests of health and safety or welfare
- not tamper with safety equipment provided by their employers for the safety of themselves or others.

The Act places responsibilities on 'persons in control of premises' to:

- ensure safe means of access and egress
- ensure that persons using premises who are not their employees are reasonably protected in respect of health and safety.

The Act also requires employers to conduct their undertaking in such a way that persons who are not their employees are not adversely affected by it in respect of health and safety.

Finally, the Act requires manufacturers and suppliers and others who design, impart, supply, erect or install any article, plant, machinery, equipment or appliance for use at work, or who manufacture, supply or import a substance for use at work, to ensure that health and safety matters are considered in respect of their product or substance.

The term 'reasonably practicable' is used in the Health and Safety at Work Etc. Act 1974 and numerous Regulations. What does it mean?

The term 'reasonably practicable' is not defined in any of the legislation in which it occurs. Only the courts can make an authoritative judgement on what is reasonably practicable.

Case law has been built up over the years and practical experience gained in interpreting the law. A common understanding of 'reasonably practicable' is:

> the risk to be weighed against the costs necessary to avert it, including time and trouble as well as financial costs.

If, compared with the costs involved of removing or reducing it, the risk is small (i.e. consequences are minor or infrequent) then the precautions need not be taken.

Any establishment of what is reasonably practicable should be made before any incident occurs.

The burden of proof in respect of what is reasonably practicable in the circumstances rests with the employer or other duty holder. They would generally need to prove why something is *not* reasonably practicable at a particular point in time.

An ability to *meet* the costs involved in mitigating the hazard and risk is *not* a factor which the employer can take into consideration when determining 'reasonably practicable'. Costs can only be considered in relation to whether it is reasonable to spend the money given the risks identified.

It is only possible to determine 'reasonably practicable' when a full, comprehensive risk assessment has been completed.

How can civil action under civil law be actioned by employees in respect of health and safety at work?

Civil action can be initiated by an employee who has suffered injury or damage to health caused by their work.

The employer may be in breach of the 'duty of care' which they owe to the employee. They may have been negligent in common law — that body of law which has been determined by case law — it has evolved rather than been set down by parliament.

Civil action may be brought on the grounds that the employer is in breach of statutory duty, i.e. has failed to follow the requirements of statutory law (e.g. Acts and Regulations). Many Acts and Regulations do not necessarily confer a right to action in civil law if statutory duty is breached but some do (e.g. Construction (Design and Management) Regulations 1994).

How quickly do employees need to bring claims to the courts in respect of civil claims?

Civil actions must commence within *three* years from the time of knowledge of the cause of action. This will usually be the date on which the employee knew or should have known that there was a significant injury and that it was caused by the employer's negligence.

Therefore, an employer would be wise to keep all records of training, risk assessments, checks, maintenance schedules, etc. for a minimum of three years as these may be needed for any defence to a claim, etc.

A civil claim will succeed if the plaintiff — the person bringing the case — can prove breach of statutory duty or the duty of care beyond 'the balance of probabilities'.

The employer may mount a number of defences to the claim, the most common of which are:

- contributory negligence — i.e. the injured employee was careless or reckless (e.g. ignored clear safety rules and procedures)
- injuries not reasonably foreseeable — i.e. the injuries were beyond normal expectation or control — the employer did not have the knowledge to foresee the risks and neither did science or experts
- voluntary assumption of risk — i.e. the employee consents to take risks as part of the job. But employers cannot rely on this defence to excuse them of fulfilling their duties under legislation — no one can contract out of their statutory duties.

What are the consequences for the employer if an employee wins their civil action?

Employers will invariably be required to pay damages or compensation. Compensation claims can run into thousands of pounds in some cases. The employer's Employer's Liability insurance will cover the cost of the claims, less any excess which the employer opts to pay.

Damages are assessed on:

- loss of earnings
- damage to any clothing, property or personal effects
- pain and suffering
- future loss of earnings
- disfigurement
- inability to lead an expected, normal personal or social life because of the injury, etc.
- medical and nursing expenses.

There will not only be the financial payout but associated bad publicity which could lead to loss of reputation.

Who enforces health and safety legislation?

There are, in the main, two organisations with powers to enforce health and safety legislation:

(1) the Health and Safety Executive and
(2) the local authorities.

The Health and Safety Executive enforces the law in the following types of work environment:

- industrial premises
- factories and manufacturing plant

- construction sites
- hospitals and nursing or medical homes
- local authority premises
- mines and quarries
- railways
- broadcasting and filming
- agricultural activities
- shipping
- airports
- universities, colleges and schools.

The local authorities, usually through their Environmental Health Departments, are responsible for:

- retail sale of goods
- warehousing of goods
- exhibitions
- office activities
- catering services
- caravan or camping sites
- consumer services provided in a shop
- baths, saunas and body treatments
- zoos and animal sanctuaries
- churches and religious buildings
- childcare businesses
- residential care.

The powers that both enforcing authorities have are the same but the way they use them may differ. Local authority inspectors may visit premises more frequently than the HSE but the HSE may be tougher on taking formal action because they have not got the resources to return to check for improvements.

What powers do enforcing authorities have to enforce health and safety legislation?

Inspectors can take action when they encounter a contravention of health and safety legislation and when they discover a situation where there is imminent risk of serious personal injury.

An inspector can also instigate legal proceedings, although, in reality, the decision to proceed to court is often taken by the enforcing authority's in-house legal team.

Inspectors may serve an *Improvement Notice* if they are of the opinion that a person:

- is contravening one or more of the relevant statutory provisions

or

- has contravened one or more of those provisions in circumstances where the contravention is likely to occur again or continue.

The inspector must be able to identify that one or more legal requirements under Acts or Regulations is being contravened, e.g. failure to complete a risk assessment, operating an unsafe system of work.

An Improvement Notice must:

- state what 'statutory provisions' are being, or have been, contravened
- state in what way the legislation has been contravened
- specify the steps which must be taken to remedy the contravention
- specify a time within which the person is required to remedy the contraventions.

The time allowed to remedy contraventions specified in an Improvement Notice must be at least 21 days. This is because there is an appeal procedure to the service of an Improvement Notice and the appeal must be brought within 21 days.

An Improvement Notice served on a company as an employer must be served on the registered office and is usually served on the Company Secretary.

If an inspector believes that health and safety issues are so appallingly managed in workplaces or premises, etc. that there is 'a risk of serious personal injury' then a *Prohibition Notice* can be served. The inspector must be able to show or prove a risk to health and safety.

Prohibition Notices can be served in *anticipation* of danger and the inspector does not have to identify specific health and safety legislation which is being, or has been, contravened.

A Prohibition Notice must:

- state the inspector's opinion that there is a risk of serious personal injury
- specify the matters which create the risk
- state whether statutory provisions are being, or have been or will be, contravened and, if so, which ones
- state that the activities described in the Notice cannot be carried on by the person on whom the Notice is served, unless the provisions listed in the Notice have been remedied.

A Prohibition Notice takes effect immediately where stated, or can be 'deferred' to a specified time.

Risks to health and safety do not need to be imminent but, usually, there must be a hazard which is likely to cause imminent risk of injury otherwise the inspector could serve an Improvement Notice.

What are the consequences of failing to comply with any of the Notices served by inspectors?

Failure to comply with either an Improvement Notice or a Prohibition Notice is an offence under the Health and Safety at Work Etc. Act 1974.

Legal proceedings are issued against the person, employer or company on whom the Notice was served and the matter will be heard in the magistrates' court in the first instance. However, serious contraventions can be passed to the Crown Court where powers of remedy are greater.

Failure to comply with an Improvement Notice carries a fine of up to £20 000 in the magistrates' court, or an unlimited fine in Crown Court, plus possible imprisonment.

Failure to comply with a Prohibition Notice carries a fine of up to £20 000 in the magistrates' court, or an unlimited fine in the Crown Court. Imprisonment of persons who contravene a Prohibition Notice is also an available option for the courts — both magistrates' and Crown Courts.

Where employers or others fail to comply with a Prohibition Notice and continue, for example, to use defective machinery, and where, as a consequence of using that defective machinery, a serious accident occurs, there will invariably be a prosecution and the courts are likely to take a serious view and hand down prison sentences.

Can an inspector serve a Statutory Notice and prosecute at the same time?

Yes. If an inspector serves an Improvement Notice or Prohibition Notice he may decide that the contravention is so blatant or so serious that immediate prosecution is warranted. The Notices ensure that unsafe conditions are remedied during the prosecution process as legal cases can take several months to come before the courts.

Employers and others do not have to be given time to remedy defects once they have been identified as many duties on employers are 'absolute' i.e. the employer 'must' do something.

Inspectors will often prosecute without initiating formal action where they believe they have given the employer ample opportunity to put right defects, e.g. they may have given advice during a routine inspection, issued informal letters, etc.

Often, when accidents are being investigated, contraventions of health and safety legislation will be viewed seriously and prosecutions will be taken.

What is the appeal process against the service of Improvement and Prohibition Notices?

Any person on whom a Notice is served can appeal against its service on the grounds that:

- the inspector wrongly interpreted the law
- the inspector exceeded his powers
- the proposed solution to remedy the default is not practicable
- the breach of law is so insignificant that the Notice should be withdrawn.

Any appeal must be lodged within 21 days of the service of the Notice.

An appeal is to an Employment Tribunal.

An Improvement Notice is suspended pending the appeal process. This means that the employer or person served with the Notice will be able to continue doing what they have been doing without changing practices or procedures.

Costs for bringing an appeal may be awarded by the Tribunal — either in favour of the enforcing authority if the appeal is dismissed or for the employer if their appeal is successful.

A Prohibition Notice may continue in force pending an Appeal unless the employer or person in receipt of the Notice requests the Employment Tribunal to suspend the Notice pending the appeal.

All Statutory Notices served by enforcing authorities should contain information on how to lodge an appeal.

On hearing the appeal, an Employment Tribunal can:

- dismiss the appeal, upholding the Notices as served
- withdraw the Notices thereby upholding the appeal

- vary the Notices in respect of the time given to complete works, i.e. the remedies listed in any schedule
- impose new remedies not contained in the Notice if these will provide the solutions necessary to comply with the law.

If remedial works cannot be completed in the time given in the Improvement Notice, can it be extended?

Yes. It is allowable for the inspector who served the Notice to extend the time limits given if works cannot be completed and a request is submitted to the enforcing authority.

It would be prudent to explain why compliance cannot be achieved, what steps have been taken in the interim and when compliance will be expected.

What are the powers of inspectors under the Health and Safety at Work Etc. Act 1974?

Enforcing authorities and their individual inspectors have wide ranging powers under the Act and, in addition to the service of notices, can:

- enter and search premises
- seize or impound articles, substances or equipment
- instruct that premises and anything in them remain undisturbed for as long as necessary while they conduct their investigation
- take measurements, photographs and recordings
- detain items for testing or analysis
- interview persons
- take samples of anything for analysis, including air samples

- require to see and copy, if necessary, any documents, records, etc. relevant to their investigation or inspection
- require that facilities are made available to them while carrying out their investigation.

If inspectors believe that they will meet resistance, they may be accompanied by a police officer. It is an offence to obstruct an inspector while they are carrying out their duties.

What is a Section 20 interview under the Health and Safety at Work Etc. Act 1974?

Under Section 20, an inspector can require anyone to answer questions as they think fit in relation to any actual or potential breach of the legislation, accident or incident investigation, etc.

A Section 20 interview is, or should be, a series of questions asked by the inspector — not a witness statement where the interviewee describes what happened.

The answers to Section 20 questions must be recorded and the interviewee must sign a declaration that they are true.

However, evidence given in a Section 20 interview is inadmissible in any proceedings subsequently taken against the person giving the interview or statement, or his or her spouse.

If an inspector is contemplating bringing criminal proceedings, he will usually opt to interview a person under the Police and Criminal Evidence Act 1984 (PACE) as the information gathered during these interviews is admissible in court. The Code of Practice on PACE interviews is strict, e.g. a caution must be given and a failure to follow the procedures could result in acquittal on grounds of technicality.

What is the law on corporate manslaughter?

There are two types of manslaughter.

- Voluntary — involving an intent to kill or do serious injury with mitigating circumstances, including provocation and diminished responsibility.
- Involuntary — for all other killings, usually sub-divided into two categories:
  - manslaughter by an unlawful or dangerous act
  - manslaughter by reckless, or possibly gross, negligence.

Companies can be prosecuted for manslaughter, but it is difficult to prove because it is not always possible to establish that the 'directing mind' of the organisation (i.e. its directors and senior managers) had sufficient knowledge about any safety contravention to be reckless, etc. Proving manslaughter charges is easier with small companies where, for instance, the Managing Director makes all the decisions and knows whether the company is complying with the law.

The Law Commission issued a report in 1996 which recommended a new central offence of 'corporate killing', committed where a company's conduct in causing the death falls far below what could reasonably be expected. It would not expect that the risk of death should be obvious or that the company should be capable of appreciating the risk. It would be sufficient to prove that the death had been caused by a company's failure in the way that its activities were managed and organised.

The Government issued a White Paper in 2000 which indicated that the recommendations of the Law Commission would be implemented into UK law. To date (2003), no new legislation has been enacted and the law on corporate manslaughter stands as currently interpreted, although in May 2003 the Home Office re-confirmed its commitment to introduce legislation when parliamentary time permits.

However, the Health and Safety Executive have issued improved guidelines on the responsibilities of directors and the courts are more active in seeing individual directors and managers prosecuted for health and safety offences.

What are the fines for offences against health and safety legislation?

Health and safety offences are usually 'triable either way' which means that they can be heard in the magistrates' courts or Crown Courts. There are some relatively minor offences which can only be heard in the magistrates' courts and some serious offences which can only be heard in the Crown Courts.

The sentencing powers of the two courts are different, with the Crown Court operating with a jury. Higher fines and imprisonment can be imposed by the Crown Courts.

Offences only triable in the magistrates' court are:

- obstructing an investigation ordered by the Health and Safety Commission
- failing to answer questions under Section 20 of the Health and Safety at Work Etc. Act 1974
- obstructing an inspector in his duties
- preventing another person from answering questions or co-operating with an inspector
- impersonating an inspector.

Offences 'triable either way' are:

- failure to comply with any or all of Sections 2–7 of the Health and Safety at Work Etc. Act 1974
- contravening Section 8 of the 1974 Act — intentionally or recklessly interfering with anything provided for safety
- levying payment on employees or others for safety equipment, etc., contrary to Section 9

- contravening any of the Health and Safety Regulations made under the 1974 Act or other enabling legislation
- contravening the powers of inspectors in relation to the seizure of articles, etc.
- contravening the provisions and requirements of Improvement and Prohibition Notices
- making false declarations, keeping false records, etc. in relation to health and safety matters.

If cases are to be brought before the magistrates' courts (summary offences) they must be brought within six months from the date the complaint is laid, i.e. lodged at the magistrates' court and summonses issued.

The magistrates' court has received guidance from the Court of Appeal that they must not hear cases where there have been serious breaches of health and safety law, fatalities or major injuries because their sentencing powers are not adequate. They should refer these for trial to the Crown Courts.

Fines can be imposed as follows.

- Breaches of Sections 2–6 of the Health and Safety at Work Etc. Act:
  - magistrates' court — a maximum of £20 000 fine for each offence
  - Crown Court — an unlimited fine for each offence.
- Breaches of Improvement and Prohibition Notices:
  - magistrates' court — a maximum fine of £20 000 or imprisonment for up to six months, or both
  - Crown Court — unlimited fine, or imprisonment for up to two years, or both.
- Breaches of health and safety regulations and other Sections of the 1974 Act:
  - magistrates' court — fines of up to £5000 per offence
  - Crown Court — unlimited fines per offence.

There are proposals to raise the level of fines and a Private Member's Bill has been tabled in 2003 for increased fines. The proposal has not yet been actioned.

The HSE issues 'guidance' and Codes of Practice. What are these and is a criminal offence committed if they are not followed?

The HSE endeavours to provide employers and others with as much information as possible on how to comply with legislation.

Guidance documents are issued on a variety of specific industries or particular processes with the purpose of:

- interpreting the law, i.e. helping people to understand what the law says
- assisting with complying with the law
- giving technical advice.

Following guidance is not compulsory on employers, and they are free to take other actions to eliminate or reduce hazards and risks.

However, if an employer does follow the guidance as laid down in the HSE documents, they will generally be doing enough to comply with the law.

Approved Codes of Practice are the other common documents issued by the HSE and these set out good practice and give advice on how to comply with the law. A Code of Practice will usually illustrate the steps which need to be taken to be able to show that 'suitable and sufficient' steps have been taken in respect of managing health and safety risks.

Approved Codes of Practice have a special legal status. If employers are prosecuted for a breach of health and safety law and it is proved that they have *not* followed the relevant provisions of the Approved Code of Practice (known as an ACOP), a court can find them at fault unless they can show that they have complied with the law in some other way.

Case study

Health and safety fines

In its third annual report *Health and Safety Offences and Penalties 2001/02*, the HSE report fines per industry sector as:

General fines levied across all sectors:	£10 million
Individual fines:	£8284 (increased from £6226)
Construction — fines per offence:	£7564
Manufacturing — fines per offence:	£9083
Extractive industries — fines per offence:	£17 550
Service sector — fines per offence:	£8832
Highest fines for health and safety offences:	£750 000
	£350 000
	£250 000
	£225 000

The HSE prosecuted 55 individuals, including cases against 31 directors.

If an employer can show that they followed the provisions of an Approved Code of Practice they will be unlikely to be prosecuted for an offence. Equally, if the employer follows the ACOP and the enforcing authority serve an Improvement or Prohibition Notice, the employer would have grounds to appeal the Notice.

What is a Safety Policy?

Under the Health and Safety at Work Etc. Act 1974, employers must produce a written Health and Safety Policy if they have five or more employees.

The policy must contain a written statement of their general policy on health and safety, the organisation of the policy and the arrangements for carrying it out.

Employees must be made aware of the Safety Policy and must be given information, instruction and training in its content, use, their responsibilities, etc.

A copy of the Safety Policy Statement must either be given to all employees or be displayed in a prominent position in the workplace.

The Safety Policy must be reviewed regularly by the employer and kept up to date to reflect changes in practices, procedures, the law, etc.

What is meant by organisational arrangements?

This section of the Safety Policy shows how the organisation will put its good intentions into practice and outlines the responsibilities for health and safety for different levels of management within the company.

An organisational section will normally include:

- health and safety objectives
- responsibilities for:

    - Managing Director or CEO
    - Operations Director
    - Safety Director
    - senior management
    - departmental heads
    - maintenance
    - employees
- training arrangements
- monitoring and review processes
- appointment of competent persons
- consultation process for health and safety
- appointment of employee representatives
- procedures for conducting risk assessments
- emergency plans.

What is meant by 'arrangements' in a Safety Policy?

The Safety Policy must either contain details of what employees and others must do in order to ensure their safety at work, or it must contain references as to where information on safe practices can be found, e.g. in the department handbook, employee induction pack, etc.

Usually, however, for ease of use and clarity, most employers will produce everything needed for the Safety Policy in one document.

The 'arrangements' section of the Safety Policy contains the details of *how* you expect your employees and others (e.g. contractors) to proceed with a task or job activity safely.

Subjects often covered under arrangements are:

- accident and incident reporting and investigations
- first aid
- risk assessments
- fire risk assessments
- manual handling

- using equipment
- electricity and gas safety
- personal protective equipment
- emergency procedures
- fire safety procedures
- training
- monitoring and review procedures
- COSHH procedures
- occupational health
- maintenance and repair
- permit to work procedures
- stress in the workplace
- violence in the workplace
- operational procedures.

A Safety Policy needs to be 'suitable and sufficient', not necessarily perfect.

If an accident happens in the workplace, the Investigating Officer (either HSE or EHO) will almost always want to see a copy of the Safety Policy and Risk Assessments. They will be looking to see if you had considered the hazard and risks of the job and implemented control measures. They will want to establish whether employees knew what to do safely and the best place to review such information is in the Safety Policy.

Prosecutions have been taken for failure to have a written Safety Policy and also for having a totally inadequate one.

The Safety Policy should be thought of as your communication tool between you, the employer and your workforce. It should be their reference guide on how you expect them to perform their job tasks safely.

3

The Management of Health and Safety at Work Regulations 1999

What is important about the Management of Health and Safety at Work Regulations 1999 (MHSW)?

The Management Regulations contain specific and goal-setting requirements for health and safety and cover all employment situations and all types of employment, self-employed persons and others whose undertaking may have a health and safety impact on people.

The Management Regulations were introduced in 1992 as part of the health and safety 'six pack' of Regulations, implemented in the UK as a result of a number of EU Directives.

The Management Regulations set out specific areas of health and safety management which employers must address and gave explicit duties to employers and others instead of the rather 'wishy-washy' implicit duties imposed at that time by the Health and Safety at Work Etc. Act 1974.

The most important aspect of the Management Regulations is the requirement for all employers and the self-employed to carry out risk assessments for all work activities undertaken by their employees or others.

The Management Regulations were amended in 1999 following various EU Directives and instructions to the UK Government that the 1992 Regulations did not fully adopt the principles of the EU Social Charter and its intent in respect of harmonising health and safety law across Europe.

The Management Regulations are the backbone of the UK's health and safety management regime and must always be used as the benchmark for health and safety compliance.

Many topic-specific Regulations will cross reference to the Management Regulations and two or more sets of Regulations will need to be read in association.

What is the requirement for risk assessments?

Regulation 3 sets down the requirements on employers to carry out risk assessments.

The duty is absolute on the employer in that the Regulation states that the employer *shall* make a suitable and sufficient risk assessment.

Regulation 3 specifically requires the employer to make a suitable and sufficient assessment of:

- risks to the health and safety of his employees to which they are exposed while they are at work, and
- risks to the health and safety of persons not in his employment arising out of or in conjunction with the conduct by him of his undertaking

for the purpose of identifying the measures the employer needs to take to comply with the requirements and prohibitions imposed on him by or under the relevant statutory provisions and by Part II of the Fire Precautions (Workplace) Regulations 1997.

Regulation 3 also requires self-employed persons to make a suitable and sufficient assessment of:

- risks to their own health and safety to which they are exposed while at work, and

- risks to the health and safety of persons not in their employment arising out of or in connection with the conduct by them of their undertaking

for the purpose of identifying measures which they need to take to comply with the requirements and prohibitions imposed on them by or under the relevant statutory provisions.

What is the legal requirement for reviewing risk assessments?

Again, Regulation 3, sub-section (3) covers the requirement for the review of a risk assessment and it states:

> any assessment such as is referred to in paragraph (1) and (2) above shall be reviewed by the employer or self-employed person who made it if:
>
> - there is reason to suspect that it is no longer valid; or
> - there has been a significant change in the matters to which it relates; and
> - where as a result of such a review changes to an assessment are necessary, the employer or self-employed person shall make them.

What is the legal requirement regarding the employment of young persons?

Before an employer employs a young person he must carry out a specific risk assessment which complies with Regulation 3 of MHSW Regulations 1999.

Employers must take particular account of:

- inexperience, lack of awareness of risks and immaturity of young persons

- the fitting out and layout of the workplace and workstation
- the nature, degree and duration of exposure to physical, biological and chemical agents
- the form, range and use of work equipment and the way in which it is handled
- the organisation of processes and activities
- the extent of health and safety training provided or to be provided to young persons
- risks from agents, processes and work as listed in a Council Directive.

Employers must record the significant findings of any assessment in writing if they have five or more employees.

What is the law on the 'principles of prevention'?

Regulation 4 of MHSW Regulations 1999 states that where an employer implements any preventative and protective measures he shall do so in accordance with the following principles:

- avoiding the risks
- evaluating the risks which cannot be avoided
- combating the risks at source
- adapting the work to the individual, especially as regards the design of workplaces, the choice of work equipment and the choice of production methods, with a view, in particular, to alleviating monotonous work and work at a pre-determined work rate and to reducing their effect on health
- adapting to technical progress
- replacing the dangerous with the non-dangerous or the less dangerous
- developing a coherent, overall prevention policy which covers technology, organisation of work, working conditions, social

relationships and the influence of factors relating to the working environment

- giving collective protective measures priority over individual protective measures
- giving appropriate instructions to employees.

A court of law would expect an employer to be able to demonstrate that he has followed the principles of prevention and protection should an employer find himself facing a prosecution for inadequate controls of risk.

What are the employer's duties in respect of health and safety arrangements?

Every employer shall make and give effect to such arrangements as are appropriate, having regard to the nature of his activities and the size of his undertaking, for the effective:

planning, organisation, control
monitoring and review

of the preventative and protective measures.

Where an employer has *five* or more employees, he shall record the arrangements.

The MHSW Regulations are quite explicit in the employer's duty to make 'arrangements'.

Employers must be able to demonstrate effective management of health and safety.

Effective management of the hazards and risks within the workplace will depend on the completion of effective risk assessments.

Employers will need to be able to demonstrate that they have a Health and Safety Management Plan.

The HSE have tried to help by producing a Guidance Book *Successful Health and Safety Management* (HS(G) 65) and they have also adopted their successful '5 Steps' strategy and have produced a free leaflet *Five Steps to Successful Health and Safety Management* (INDG 132L).

What are the five steps to successful health and safety management?

The five steps are:

Step 1: Set your policy
Step 2: Organise your staff
Step 3: Plan and set standards
Step 4: Measure your performance
Step 5: Audit and review

Step 1: Set your policy

- Do you have a clear Health and Safety Policy?
- Is it written down?
- Is it up to date?
- Does it specify who is responsible for what and who has overall safety responsibility?
- Does it give responsibilities to directors and is there a clear commitment that the health and safety culture starts at the top?
- Does it specify arrangements for carrying out risk assessments, identifying hazards, implementation of control measures?
- Does it name competent persons?
- Does it state how health and safety matters will be communicated throughout the organisation?
- Does it have safety objectives?

- Has it had a beneficial effect on the business?
- Does it imply a pro-active safety culture within the organisation?

Step 2: Organise your staff

- Have you involved your staff in your health and safety policy?
- Are you 'walking the talk'?
- Is there a health and safety culture within the organisation?
- Have you adopted the 'four Cs':
    - competence
    - control
    - co-operation
    - communication?
- Are your staff and others competent to do their jobs safely?
- Are they properly trained and informed?
- Are there competent people around to help and guide them?
- Have you designated key people responsible for safety in each of the business areas?
- Do employees know how they will be supervised in respect of health and safety?
- Do employees know to whom to report faults and hazards and what will be done to rectify them?

Step 3: Plan and set standards

- Have you set objectives with your employees?
- Have you reviewed accident records to see what general standards of health and safety you have in place?
- Have you set targets and benchmarks?
- Is there a purchasing policy regarding the safety standards of products etc.?
- Are there procedures for approving contractors?
- Have safe systems of work been identified?

- Have risk assessments been completed?
- Is the hierarchy of risk control followed?
- Have hazards to persons other than employees been assessed?
- Has a training plan and policy been developed and are there minimum levels of training for all employees?
- Are targets and objectives measurable, achievable and realistic?
- Is there an Emergency Plan in place?
- Have fire safety procedures been completed?
- Is there a 'zero tolerance' policy on accidents?
- Has the safety consultation process with employees been established?
- Is there a culture of continuous improvement in respect of health and safety?

Step 4: Measure your performance

- Have you or can you measure with regard to your health and safety performance:
  - where you are
  - where you want to be
  - where the difference is
  - why?
- Are you practising active monitoring or simply reacting when things go wrong?
- Is there a culture of recording near misses or do you wait for accidents to happen?
- Can you benchmark how you perform against another department, company or other organisation?
- Do you know how well you are really doing or does it just 'look good on paper'?
- Are there ongoing accident and incident records and trend analysis?
- Is the effectiveness of training measured — do you assess learning outcomes?

- Is there good legal compliance with health and safety law? Are you up to date with legal changes?

Step 5: Audit and review

- Are you regularly checking that the business is safe and minimising risks to health and safety?
- Is there a formal audit review process?
- Are risk assessments reviewed pro-actively?
- Are accidents investigated and processes changed as a result?
- Is there a formal audit process?
- Is the audit process independent?
- Do staff get involved?
- Do you share the findings of the reviews?
- Is the Board kept up to date?
- Is there a Health and Safety Committee?
- Is your business genuinely improving in respect of managing health and safety?

Managing health and safety is no different to managing any aspect of a business and it should be considered to be just as important as finance, sales, etc.

The likelihood of new corporate manslaughter legislation should focus the minds of all employers on addressing the issue of a thorough and comprehensive health and safety management system.

What are an employer's duties in respect of health surveillance?

Every employer shall ensure that his employees are provided with such health surveillance as is appropriate, having regard to the risks to their health and safety which are identified by the risk assessment.

Health surveillance is required under the COSHH Regulations 2002 and also under the Management of Health and Safety at Work Regulations 1999.

If a risk assessment shows:

- that there is an identifiable disease or adverse health condition related to the work concerned
- that valid techniques are available to detect indications of the disease or condition
- that there is a reasonable likelihood that the disease or condition may occur under the particular conditions of work
- that surveillance is likely to further protect the health and safety of employees

then health surveillance should be introduced.

A competent person must determine the level and frequency of health surveillance.

The primary benefit of health surveillance is to detect adverse conditions at an early stage, thereby enabling further harm to be prevented.

It may be necessary to consult with the Employment Medical Advisory Service if, as an employer, you do not have a competent person to assist with health surveillance assessments.

There are some specific Regulations which require an employer to assess the need for, and then offer, health surveillance and these need to be considered in addition to the general duties under MHSW:

- Control of Substances Hazardous to Health Regulations 2002
- Control of Lead at Work Regulations 2002
- Control of Asbestos at Work Regulations 2002
- Noise at Work Regulations 1989.

Under the above Regulations, medical examinations may also be required.

As an employer, can I deal with all health and safety matters myself?

Yes, provided that you could convince the enforcing authorities and probably the courts that you were competent to deal with health and safety matters.

Regulation 7 of the Management of Health and Safety at Work Regulations 1999 requires an employer to appoint one or more 'competent' persons to assist him in undertaking the measures he needs to take to comply with the requirements and prohibitions imposed on him by virtue of the relevant statutory provisions and by Part II of the Fire Precautions (Workplace) Regulations 1997.

'Competency' is described as having sufficient training and experience or knowledge and other qualities to enable him properly to assist in undertaking the measures needed to comply with statutory duties.

Any or all competent persons appointed must co-operate with one another.

The employer must ensure that any competent person is provided with relevant information about hazards and risks, etc.

Where practicable and should a competent person be available, they should be appointed from within the workforce and should not be external appointments unless specialised knowledge and experience is required.

What are the requirements regarding 'serious and imminent' danger?

Employers must adopt appropriate procedures for dealing with situations which cause serious and imminent danger to persons at work.

Competent persons must be nominated in sufficient numbers to implement the emergency plans; in particular, any evacuation of people from the workplace.

Employees must be made aware of the hazards which could cause an imminent or serious situation to arise and must be informed of the procedures to be followed to protect them from the danger.

In any situation which creates serious or imminent danger to employees, there must be a procedure which will enable them to evacuate the area safely and to proceed to a place of safety. It should be accepted that work processes will stop immediately.

Emergency procedures must be written down and available for all employees and others.

A risk assessment should determine what emergency procedures will be necessary and the employer should consider more than just fire and bomb procedures.

Examples of emergency procedures will be:

- fire
- bomb
- explosion
- chemical release
- flood
- toxic gas release, fumes, etc.
- unexpected shut-down of exhaust ventilation which could cause toxic fumes, dust, etc. to build up
- terrorist attack
- radiation leak
- biological agent release.

Employers need to consider the possibilities of the above and plan for the event by considering what they would do, who would do what, how people would get out, where they would go, how they would be accounted for, etc.

What is required regarding 'contact with emergency services'?

Employers need to assess the need for, and likelihood of, contacting the emergency services.

Often, emergency services are not familiar with the hazards associated with employers' undertakings and emergency personnel can be placed at great health and safety risk.

Regulation 9 of MHSW Regulations 1999 requires a formal arrangement to be made so that the emergency services have some knowledge of any special safety hazards, or so that they will be informed if the employer believes that there is a risk of a 'serious and imminent danger' occurrence.

Records should be kept of any contact, meetings, reviews, etc. with the emergency services.

It may be appropriate to send a copy of the Emergency Plan to the authorities for their fire records.

For many employers, it will be sufficient to ensure that employees know the emergency services' phone number and know how to contact them for assistance.

What is the legal duty under the Management of Health and Safety at Work Regulations 1999 for information to be given to employees?

Regulation 10 governs the provision of information to employees and requires that employers provide employees with 'comprehensible and relevant' information on:

- the risks to their health identified by the risk assessments
- preventative and protective measures
- emergency procedures
- fire safety arrangements
- the identity of nominated persons who will take charge of emergency situations, e.g. Fire Wardens
- any risks notified to the employer as being present on another employer's premises and to which the employee may be exposed.

When an employer proposes to employ a young person he shall inform that person's parents or guardians of the risk assessment, control measures, etc. to be adopted so as to ensure that the young person is kept safe while at work.

What happens in relation to health and safety when two or more employers share a workplace?

Health and safety law requires that two or more employers who share a workplace must co-operate with one another in respect of health and safety matters and where necessary, appoint a co-ordinator for health and safety.

Each employer must take reasonable steps to reduce risks both to his own employees and to others and must not put other people at undue risk.

Often, a multi-occupied building will have a Managing Agent and it would be the responsibility of this person, as the person in control of the premises, to co-ordinate emergency procedures, etc. and to undertake risk assessments of the 'common parts'.

Employers have a duty to inform those who need to know about the hazards and risks associated with their undertaking. Every employer must consider the risks posed by another employer when determining his own risk assessments.

It is good practice to swap Health and Safety Policies with other occupiers so that each is fully aware of the hazards and risks posed by each employer and the control measures each has in place.

Emergency procedures (e.g. evacuation) need to be co-ordinated so that there is a unified response to any alarm, etc.

Where employees are working in a host employer's premises (e.g. maintenance workers), the host employer must ensure that they provide the visiting employees with relevant information on hazards, risks, control measures and emergency procedures.

In particular, the names of nominated persons in respect of emergencies should be made available to visiting employees, e.g. Fire Wardens, first aiders.

Common practices for sharing information on safety matters for visiting employees include:

- signing in and induction briefing
- contractors' handbook
- pre-approval of contractors
- joint training sessions
- permit to work systems.

What are an employer's duties regarding the provision of training for his employees?

Regulation 13 of the Management of Health and Safety at Work Regulations 1999 requires employers to provide their employees with adequate training in respect of health and safety:

- on being recruited
- on being exposed to new or increased risks due to:
  - being transferred to another department or section
  - the provision of new equipment
  - the change of existing equipment
  - the introduction of new technology
  - the introduction of a new system of work.

Training shall be repeated periodically and shall take place during working hours. It shall also be free of charge.

The provision of information, instruction and training must be comprehensible and understood by all employees.

For the purposes of health and safety, temporary workers must be treated as permanent employees and must also receive information, instruction and training.

4

Undertaking risk assessments

Who should undertake a risk assessment?

Regulation 3 of the Management of Health and Safety at Work Regulations 1999 states that:

> every employer shall make a suitable and sufficient assessment of:
> (a) the risks to the health and safety of his employees to which they are exposed whilst they are at work; and
> (b) the risks to the health and safety of persons not in his employment arising out or in connection with the conduct by him of his undertaking,
>
> for the purposes of identifying the measures he needs to take to comply with the requirements or prohibitions imposed upon him by or under the relevant Statutory Provisions and by Part II of the Fire Precautions (Workplace) Regulations 1997.

So, the law requires an *employer* to carry out a risk assessment.

Under the terms of the organisation's Safety Policy, the employer can delegate the responsibility for undertaking risk assessments to others, e.g. Safety Officers, Departmental Managers, etc.

With the proviso that the employer carries ultimate responsibility for the risk assessment, it can be conducted by anyone authorised to do so.

In many organisations, the people best qualified to carry out a risk assessment on job tasks undertaken are the employees. This is because they are familiar with the hazards and risks of what they do and they know the actual way they carry out the tasks in practice, as opposed to the 'theoretical way'.

The law requires that anyone involved in health and safety matters for an employer, including the employers themselves, must be competent.

Regulation 7 of the MHSW requires employers to appoint competent persons to assist in health and safety matters. This will include a competent person being appointed to undertake risk assessments.

Regulation 7 defines a person as being competent if they have sufficient information, knowledge and experience to enable them to properly assist the employer in discharging his responsibilities.

Does absolutely every single job activity require a risk assessment?

No, although it sometimes feels like that!

The requirement of the Management of Health and Safety at Work Regulations 1999 is as follows:

> Every employer shall make a suitable and sufficient assessment of:
> (a) the risks to health and safety of his employees to which they are exposed while they are at work, and
> (b) the risks to the health and safety of persons not in his employment arising out of or in connection with the conduct by him of his undertaking,
>
> for the purpose of identifying the measures he needs to take to comply with the requirements and/or prohibitions imposed upon him under the relevant statutory provisions and by Part II of the Fire Precautions (Workplace) Regulations 1997 (amended).

If a work activity does not pose any health and safety risks then there is no need to carry out a risk assessment, although a risk assessment of sorts will be carried out in order to establish that the job task has no hazards or risks attached to it.

Increasingly, however, it is best practice to undertake risk assessments for all job tasks because, even though the statutory laws may not require them, the need to provide a duty of care under civil law makes risk assessments a valuable defence tool.

Is there a standard format for a risk assessment?

No, mainly because risk assessments are individual to job tasks and need to be 'site specific'!

The HSE does publish guidance on how to complete risk assessments and they include a risk assessment template.

There is no right or wrong way to complete a risk assessment. The law requires that it is 'suitable and sufficient'.

A risk assessment must contain suitable information to be useful to an employee to understand what hazards they may be exposed to when carrying out the task.

Generally, any format that includes the following will be suitable:

- description of the job task
- location of activity
- who will carry it out
- who else might be affected by the task
- what are the hazards identified
- what could go wrong
- what might the injuries be and how severe might they be
- how likely are the risks
- what can be done to reduce or eliminate the hazards
- what information do employees or others need to work safely
- when might the risk assessment be reviewed?

What do the terms 'hazard' and 'risk' mean?

A hazard is something with the potential to cause harm.

The risk is the likelihood that the potential harm from the hazard will be realised.

The extent of the risk will depend on:

- the likelihood of the harm occurring
- the potential severity of that harm (resultant injury or adverse health effect)
- the number of people who might be affected — several people, vast groups, communities at large (e.g. from chemical releases).

What does 'suitable and sufficient' mean in respect of risk assessment?

The phrase 'suitable and sufficient' is not defined in the Regulations nor within the Health and Safety at Work Etc. Act 1974.

The Approved Code of Practice on the Management of Health and Safety at Work Regulations 1999 states that:

> the level of risk arising from the work activity should determine the degree of sophistication of the risk assessment.

Insignificant risks can generally be ignored, as can routine activities associated with life in general.

Risk assessments are expected to be proportionate to the hazards and risks identified.

Enforcement Officers do not expect to see huge volumes of paperwork — the simpler the risk assessment and the clearer the information, the easier the employee will find following safe procedures.

When should risk assessments be reviewed?

The Management of Health and Safety at Work Regulations 1999 requires a risk assessment to be reviewed if:

- there is reason to suggest that it is no longer relevant or valid
- there has been significant change in the matters to which it relates.

If changes to the risk assessment are required, the employer has a duty to make the changes and re-issue the risk assessment.

Employers are not expected to anticipate risks that are not foreseeable.

If events happen, however, which alter information available or the perception of risk, the employer will be expected to respond to the new information and assess the hazards and risks in the light of the increased knowledge.

Accidents and near misses should be investigated, as these incidents will indicate whether more knowledge is available on the hazard or risk associated with the job. The risk assessment may need to be reviewed because:

- something previously unforeseen has occurred
- the risk of something happening or the consequences of the event may be greater than expected or anticipated
- precautions prove less effective than anticipated.

New equipment, new working environment, new materials, different systems of work, etc. will all require existing risk assessments to be reviewed.

Which people must be considered as being exposed to the risks from work activities?

The following people must be taken into consideration:

- employees of the employer
- young workers and those on work experience
- new and expectant mothers
- cleaners — whether contract or in-house
- visitors to the premises
- maintenance workers — both contract and in-house
- members of the public
- employees of other employers with whom you share the building or premises
- delivery drivers
- sales representatives
- people with disabilities —extra control measures may be needed to protect them from risk
- peripatetic workers — those working away from the office, usually visiting other workplaces or people's homes, e.g. midwives
- volunteers.

Must risk assessments be categorised into high, medium or low risks?

Not necessarily by law, but it is good practice to identify the extent of harm that an employee or other person could be exposed to.

Even after all precautions have been taken, some risk (i.e. potential cause of injury or ill health) may remain. This is often referred to as *residual risk.*

Residual risk is either 'high, medium or low', or very likely, probable or unlikely.

Some risk management approaches allocate numerical scores to various types of risk and the severity of those risks. By multiplying one score by the other they arrive at a 'risk rating'. Scores above a set target become unacceptable and measures must be put in place to reduce the risks.

What are some of the common control measures which can be put in place to reduce the risks from job activities?

The aim of risk assessment is to reduce the residual risk associated with a task to as small a level as possible.

First, try to eliminate the hazard altogether — why do something or use something if you do not have to?

Where the hazard cannot be eliminated it must be reduced to acceptable levels by implementing *control measures*.

A common approach is by following the 'hierarchy of risk control', namely:

- try a less risky option, i.e. substitute something less hazardous
- prevent access to the hazard, e.g. by guarding
- organise work to prevent exposure to the hazard
- issue personal protective equipment
- protect the workforce as a whole, e.g. through exhaust ventilation
- provide welfare facilities to aid removal of contamination, to take rest breaks, etc.
- provide first aid facilities.

What will an Enforcement Officer expect from my risk assessments?

Enforcement Officers will want to see that you:

- have completed risk assessments
- have considered site-specific issues
- have completed comprehensive checks of the workplace
- have involved workers
- have considered the hazards and risks to others
- have dealt with immediate hazards to reduce risks

- have in place a system for reviewing risk assessments
- keep suitable records
- have provided information, instruction and training to your employees
- have introduced suitable and sufficient measures.

Site-specific risk assessments are probably the most important aspect. EHOs and HSE Inspectors are not keen on generic risk assessments unless steps have been taken to ensure that any special site hazards and precautions have been added to the risk assessment.

To Enforcement Officers, risk assessment is not a paper exercise to be pulled off the shelf in a manual. It is a pro-active approach by an employer to consider what could harm his employees and others and what measures he intends to take to reduce the risks of injury and ill health.

What is meant by the term 'control measure'?

A control measure is the term used for the precautions deemed to be necessary in order to reduce the consequences of the hazard and risk, i.e. to reduce the risk to an acceptable level.

A control measure could be:

- physical, e.g. fixing a guard to a machine
- substitution — for a less hazardous substance
- establishing a system of work
- providing personal protective equipment
- separation of workers from an environment or machine
- environmental controls, e.g. ventilation.

The risk assessment identifies what control measures or precautions are necessary in order to manage the risks that have been identified from the hazard being assessed.

What is the 'hierarchy of risk control'?

When a hazard has been identified, the most effective way of reducing the effect of the hazard, or the risk, is by eliminating the hazard completely. Unfortunately, this is not always possible and a stayed approach to controlling the risk has to be adopted.

This stayed approach to controlling risk is called the 'hierarchy of risk control'.

The first and most effective stage of dealing with a hazard and its risk is to eliminate the hazard completely. No hazards = no risks.

The second stage, if the first cannot be achieved, is to substitute the identified hazard for a less harmful hazard; for example, why use a substance which can cause cancer when an alternative is on the market which may only be a minor skin irritant; why carry 50 kg bags of cement when 25 kg ones are available and easier to carry?

The third stage to adopt if a suitable alternative cannot be found and the original hazard has to remain is to protect the workforce as a whole from the hazard, e.g. increase the workshop ventilation so that the hazard is tackled at source.

Next, if stage three fails, provide all employees with individual worker protective clothing and equipment so that their own individual health is protected, e.g. masks, goggles, local exhaust ventilation.

Finally, monitor and review the controls that you have put in place to make sure that they are effective, otherwise it will be back to stage one — eliminating the hazard.

What is the common approach to risk assessment?

The Health and Safety Executive have pioneered a *five steps* approach to risk assessment and they have summarised this approach in their free leaflet *5 Steps to Risk Assessment* (INDG 163).

The five steps to risk assessments are:

Step 1: Look for the hazards
Step 2: Decide who might be harmed and how

Step 3: Evaluate the risks and decide whether the existing precautions are adequate or whether more should be done
Step 4: Records your findings
Step 5: Review the assessment and revise it whenever necessary.

What needs to be considered for each of the 'five steps'?

Firstly, risk assessment should not be overcomplicated. Remember that risk assessment is a careful examination of what, in your work, could cause harm to people and an assessment of what you need to do to prevent harm to people. You may already have taken enough precautions to protect them or you may need to implement some more.

You will need to decide whether a hazard is significant and whether you have covered it by satisfactory precautions so that the risk is small.

Step 1: Identify the hazards

Walk around the workplace and look out for anything which could cause harm. Take a fresh view and do not make assumptions about things.

Concentrate on identifying the serious hazards which could cause major harm to people, e.g. cause them to have a significant injury or which may affect several people.

Talk to your employees and ask them what they think of as hazards, ask them how they really do the job and whether they follow the rules or do it a little differently.

Review any records that you have such as accidents and incidents. It is a good idea to keep 'near miss' records because these should tell you that something is not working but that things have not yet become sufficiently serious to cause an accident.

What type of accidents are happening and why? Are people off sick for periods of time with back injuries, for instance.

Check manufacturers' operating procedures, manuals and instructions.

Check what substances are being used and what substances are being produced by the works process itself, e.g. dust.

Common hazards include the following:

- equipment — how it is used, guards, controls, noise
- work processes — how things are done, systems to be followed
- environmental conditions — condition of floors, heating, ventilation, etc.
- materials in use — chemicals, gases, substances.

Step 2: Decide who might be harmed and how

Consider absolutely everybody who could be harmed by the hazards you have identified.

Remember to include:

- all employees
- agency staff
- self-employed people
- visitors, public, etc.
- contractors
- cleaners
- delivery personnel
- maintenance workers.

How might they be affected by the hazard? Are they likely to incur a serious injury, or none at all? Could they be harmed because they were in the vicinity of the hazard or because they have to undertake the job task?

Step 3: Evaluate the risks and decide whether existing precautions are adequate or if more should be done

Consider how likely it is that each hazard you have identified could cause harm.

Then consider what you are currently doing to reduce the potential for harm?

Are you doing enough?

Have you done all the things that the law requires you to do?

Are you following good practice and industry standards? Are manufacturers' instructions being adhered to?

If, for instance, you use dangerous machinery, do you have guards in place so that the dangerous parts cannot be accessed?

What more could be done to reduce the risks to as small a level as possible?

If you are not satisfied that all risks are as small as possible then more needs to be done.

Draw up an Action Plan so that you work on eliminating the highest risks first, or those which could harm most people.

If you need to provide more control measures or precautions, consider the 'hierarchy of risk control':

- eliminate the hazard
- substitute for a less risky option
- prevent access to the hazard
- organise the work to reduce exposure to the hazard
- issue personal protective equipment
- provide welfare facilities, e.g. washing facilities, first aid, etc.

Consider all types of work environments in which employees work and whether they work in someone else's premises. An employer must assess the hazards and risks to his employees and, even though you may not know what they are, you are responsible if your employee works somewhere else. So, ask the owner of that building or business what *their* hazards and risks are and assess those in relation to the job you expect your employee to do.

Example

You may have identified the hazard of working at heights — there is definitely a potential harm to people from falling. You may have control measures in place because you have provided scaffolding and handrails and toeboards as required by the Construction (Health, Safety and Welfare) Regulations 1996. But have you done enough? Have you considered whether your employees or others *need* to work at height in the first place? Could they have a safer means of access to the place of work than the scaffold? In itself, a scaffold is dangerous or hazardous. Could mobile elevating work platforms be used?

Persons in control of premises have duties under health and safety even if they are not employees and the Management of Health and Safety at Work Regulations 1999 cover the co-operation of employers and others where there is a multi-occupied site.

Step 4: Record your findings

If you have five or more employees, you need to record the *significant findings* of your risk assessment process in writing.

Small companies with fewer than five employees are exempt, although it is always good practice to keep some records, as you never know when you might want to prove what you have done.

As stated earlier, there is no standard risk assessment form or template for records.

You will need to decide what type of form is appropriate for record keeping. The HSE give some guidance on a simple layout for a risk assessment but there is no right or wrong way to go about it.

Remember that your risk assessment must be 'suitable and sufficient' — this does *not* mean perfect!

The records that you keep need to show that you have considered the hazards, identified the people at risk, determined and actioned the control measures necessary to reduce the risks and considered when the risk assessment needs to be reviewed.

Step 5: Review and revise the risk assessment

Risk assessment is not a 'once and for all' exercise. Hazards in any workplace may change, the circumstances in which those hazards occur may change, the people may change, and the materials and equipment may change!

So, a risk assessment needs to be constantly reviewed to see if it is still relevant.

Your original risk assessment process should identify when and if hazards and risks will change and should indicate a regular review period.

If things stay fairly static, a risk assessment with a low residual risk may only need to be reviewed annually. But a works process or job task with a high residual risk which relies on effective control measures to make the risk tolerable will need to be reviewed much more regularly.

What are site-specific risk assessments?

Site-specific risk assessments review the hazards and risks associated with an actual job on a specific site or within specific premises.

The law is concerned about what may *actually* happen to an employee or other person while they are at work, or affected by work activities, not what could happen because the employer has brainstormed every conceivable hazard in every conceivable location.

Hazards may be quite common across a range of work activities and there is a tendency for employers to produce 'generic risk assessments'. These are records of common hazards and risk but they do not address the actual work environment.

Generic risk assessments do have a valuable part to play in the process of risk assessment but they need to be reviewed in the light of what actually happens in the workplace.

Are generic risk assessments acceptable under health and safety law?

There is no law against a generic risk assessment produced by, for example, a trade body being used as the employer's risk assessment.

However, the question to consider is whether the generic risk assessment is 'suitable and sufficient' for the purposes of the Management of Health and Safety at Work Regulations 1999.

Prosecutions can be taken by the enforcing authorities for inadequate risk assessments and this is just as serious an offence as having *no* risk assessments.

An enforcing authority is more likely to serve an Improvement Notice on the employer for failing to have a suitable and sufficient risk assessment and will require improvements to be made. Alternatively, if the inspector is concerned about the risk from the job task or activity, he could serve a Prohibition Notice under the Health and Safety at Work Etc. Act 1974.

As an employer, I have completed my risk assessments, recorded the significant findings in writing and have a Risk Assessment Manual. What else do I need to do?

The law requires you to tell your employees and others about the findings of your risk assessment exercise.

Employees must know what hazards they are exposed to during their working day and must be advised of the precautions or controls which have been implemented to help reduce the risk of harm.

Employees could be given copies of the risk assessments individually as part of their employee handbook, or they could be given a formal training session which identifies the hazards and risks and trains them in how to operate or follow the control measures.

Information could be displayed on company notice boards or adjacent to the work areas.

As long as employees are kept informed, the method by which it is done is left to the employer.

Every employer is under a duty to ensure that employees receive information, instruction and training in respect of any matter which may affect their health and safety while at work.

5

Control of substances hazardous to health and dangerous explosive substances

What are the COSHH Regulations 2002?

The Control of Substances Hazardous to Health Regulations 2002 are known as COSHH.

The Regulations set out the duties that employers have to their employees and others to protect them from exposure to and harm from hazardous substances.

The 2002 Regulations came into force in November 2002 and replace all earlier sets of Regulations, i.e. 1988, 1994 and 1999.

The Regulations were amended in March 2003 to address further issues in respect of carcinogens.

What does COSHH require?

Five basic principles of occupational hygiene underpin the COSHH Regulations:

(1) identify the hazardous substance, identify how it is to be used, assess the risk to health, precautions and health risks arising from that substance

(2) if the substance is harmful, wherever possible, substitute a less harmful substance
(3) introduce appropriate measures to prevent or control risks and ensure that control measures are used, that any protective equipment is properly maintained and that any safety procedures are observed
(4) where necessary, monitor the exposure of employees and introduce an appropriate form of *surveillance of their health*
(5) inform, instruct and train employees in the risks to their health and safety and the precautions that need to be taken.

What substances are covered by the COSHH Regulations?

The Regulations cover a wide range of substances and include those which are very toxic, harmful, corrosive, irritant or biological. These could include cleaning materials for floors, toilets, drains, glasswasher and dishwasher detergents, pest control materials, dusts, fumes, solvents, building products, oils, etc.

The COSHH legislations will apply to hazardous substances no matter how large or small the quantity. The overriding principle is that if a substance is a hazard to health, it *must* be assessed.

All of these substances are safe when properly used, but the use of each must be assessed. Employees must be made aware of any hazards, the precautions necessary and trained how to use the substances correctly.

What changes did the 2002 COSHH Regulations introduce?

The 2002 Regulations did not fundamentally change employers' duties to ensure that employees and others are not exposed to the harmful effects of hazardous substances.

The Regulations generally have made changes as follows:

- to numerous definitions within the Regulations, e.g. biological agents, inhalable and respiratable dust
- COSHH assessments under Regulation 6 have been amended to:
    - require that the steps identified by the assessment as necessary to meet the requirements of the Regulations are implemented
    - the assessment is to consider
        - — the hazardous properties of the substance
        - — information on health effects provided by the supplier, including information contained in the Safety Data Sheets
        - — the level, type and duration of exposure
        - — the circumstances of the work, including the amount of substance involved
        - — activities, such as maintenance, where there is potential for a high level of exposure
        - — any relevant occupational exposure limit or standard, maximum exposure limit or similar occupational exposure limit
        - — the effect of preventative and control measures which have been or will be taken to comply with Regulation 7
        - — the results of relevant health surveillance
        - — the results of any monitoring of exposure
        - — the risks of exposure to more than one substance, i.e. the 'cocktail' effect
        - — the approved classification of any biological agent
        - — such additional information as the employer may need to complete the assessment
- the assessment is to be reviewed if the results of monitoring show it to be necessary

- employers who employ five or more employees are to record the significant findings of the assessment as soon as is practicable after the risk assessment is made and steps are to be taken to implement control measures
- a specific requirement is introduced under Regulation 7 to substitute a substance or process if this eliminates or reduces risks to health
- control measures are listed in order of priority in Regulation 7
- biological agents are now covered in the body of the Regulations
- all control measures are to be kept clean
- new provisions regarding employee monitoring, the keeping of records and health surveillance have been introduced
- information, instruction and training requirements have been extended to include details on occupational exposure limits, access to relevant safety data sheets, exposure and health risks of the substance, significant findings of the COSHH assessment, results of health surveillance, control measures to be implemented, etc.

Duties extend to training persons other than the employer's employees if those people are so exposed to the risks from hazardous substances.

How do dangerous chemical products get into the body?

There are three main ways in which products get into the body: through ingestion, through the skin or through inhalation. The form of the product plays an important role. The more finely divided a product is, the more easily it is absorbed (generally the smaller the particles, the more dangerous they are). Solids for example, may be in the form of powder and liquids in the form of an aerosol.

Absorption is dependent on many factors, including the state of subdivision of the product (i.e. the smallness of the particles), its concentration, the length of exposure, the use of protective equipment, its solubility in fat, etc.

Digestive route (via the mouth)

Entry via the digestive route (or ingestion) is usually accidental or the result of carelessness, for example:

- through transferring a product from one container to another by sucking it up through a pipette, or through a product having been stored in a food and drink container
- through eating, smoking, drinking, etc. after having handled a dangerous product and not having washed hands.

Percutaneous route (entry via the skin)

Certain products, such as irritant and corrosive products act locally at the place where they come into contact with the skin, the mucous membrane or the eyes.

Others, which are soluble in fat, not only act on the skin but also penetrate it and spread throughout the body where they can cause various disorders. This is the case with solvents, which degrease the skin, but which can also damage the liver, nervous system or kidneys. Benzene can damage the bone marrow. Motor fuel (which has a relatively high benzene content) should not be used to wash hands.

Small cuts and grazes provide an easy route for dangerous chemicals.

Respiratory route (entry via the lungs)

This is the most common entry route at work, as pollutants can be present in the atmosphere. They then enter the lungs with the air we breathe. This can occur when handling solvents, paints or glues, stripping leaded paint with a blow torch or welding, for example.

Once inhaled into the lungs, these chemicals enter the bloodstream and can cause damage not only to the respiratory system but also to the rest of the body.

A chemical which enters via any of these routes can be transported to other parts of the body in the bloodstream and can cause damage to other organs.

What are Occupational Exposure Limits or Standards (OEL/OES)?

For a number of commonly used hazardous substances, the Health and Safety Commission has assigned occupational exposure limits (or standards) to help define what is adequate control.

Occupational Exposure Limits are set at levels which will not damage the health of employees exposed to the substance by inhalation, day after day.

Where a substance has an OEL, the exposure of employees to the substance must legally be reduced to the OEL level.

What are maximum exposure limits?

Maximum exposure limits are set for substances which can cause the maximum amount of health damage. These substances usually cause life-threatening illnesses such as cancer, asthma, severe industrial dermatitis, respiratory conditions, etc.

Substances which have an MEL must be used only if there is no alternative and exposure must not exceed the stated limit over the given exposure time — usually no more than ten minutes.

Employers should avoid the use of all substances with an MEL — find an alternative.

What is health surveillance?

Health surveillance is required under certain circumstances and requires employers to assess the health of their employees regularly. If employees are exposed, for instance, to a substance which causes skin irritation, then it may be necessary to check the condition of hands and arms by visual examination from time to time.

Health surveillance allows an employer the opportunity to monitor the effectiveness of the control measures in place.

If employees are exposed to breathing in fumes or dust, then routine lung tests or blood tests can be used.

Health surveillance can be carried out by a medical doctor or occupational nurse, or an employer can carry out simple assessments and refer to experts for advice.

Recognising hazardous substances

See table overleaf.

Symbol	Meaning	Description of risks
	Toxic (T) Very toxic (T+)	Toxic and harmful substances and preparations posing a danger to health, even in small amounts. If very small amounts have an effect on health the product is identified by the toxic symbol.
	Harmful (Xn)	These products enter the organism through inhalation, ingestion or the skin.
	Highly flammable (F) Extremely flammable (F+)	(F) Highly flammable products ignite in the presence of a flame, a source of heat (e.g. a hot surface) or a spark. (F+) Extremely flammable products can readily be ignited by an energy source (flame, spark, etc.) even at temperatures below 0ºC.
	Oxidising (O)	Combustion requires a combustible material, oxygen and a source of ignition; it is greatly accelerated in the presence of an oxidising product (a substance rich in oxygen).

	Corrosive (C)	Corrosive substances seriously damage living tissue and also attack other materials. The reaction may be due to the presence of water or humidity.
	Irritant (Xi)	Repeated contact with irritant products causes inflammation of the skin and mucous membranes, etc.
	Explosive (E)	An explosion is an extremely rapid combustion. It depends on the characteristics of the product, the temperature (source of heat), contact with other products (reaction), shocks or friction.
	Dangerous for the environment (<<N)	Substances which are highly toxic for aquatic organisms, toxic for fauna, dangerous for the ozone layer.

How might the risk of accidents from hazardous substances be reduced?

- Check that packages and containers are in good condition, so as to avoid leaks. Make sure the gases, fumes, vapours or dusts are extracted at their point of origin. Wear a respirator if necessary. Watch out for possible sources of fire.
- Keep dangerous products only in appropriate containers, property labelled. Never transfer them into bottles such as lemonade or beer bottles, or other food containers. This type of practice causes serious accidents every year. Dangerous products should preferably be kept locked away when not in use.
- Avoid contact with the mouth. Do not eat, drink or smoke when using dangerous substances or when in a place where they are used.

What steps can be followed when establishing whether to substitute a less harmful substance for a hazardous one?

It is often difficult to decide whether to substitute a substance used at work because it is not always clear whether substances actually are less harmful and, often, the attitude that 'we have always used this' is common place and people's desire to embrace change is limited.

However, the HSE advocate a *seven step* approach to harmful substance substitution and they advocate that a logical approach will enable all aspects to be considered thereby ensuring that a correct decision is reached.

Companies may wish to substitute hazardous substances for less harmful substances for health and safety reasons, but the decisions are also valid for finding substances that are less harmful to the environment.

The seven steps

Step 1: Decide whether the current substance or process is a hazard. Is there a significant risk involved in storing, using or disposing of the substance?
Step 2: Identify the alternatives.
Step 3: Think about what could happen if you use the alternatives.
Step 4: Compare the alternatives with each other and with the process or substance you are using at the moment.
Step 5: Decide whether to substitute.
Step 6: Introduce the substitute.
Step 7: Assess how the substitute is working.

Step 1

You should already have completed a COSHH assessment for the substances under consideration. If you have assessed that you are using substances that have a high risk of injury or disease, they should be substituted, e.g. any substance which is known to be carcinogenic.

Consider also any risk of fire, explosion or environmental risk.

Step 2

Identifying alternatives needs to be wide ranging. The first question is 'do you need to do the job at all?'.

Alternatives may not be other substances but may be a change of process, e.g. using drain rods to clear blockages rather than using sulphuric acid.

Consult with the users and find out in exactly what quantities and how often they use the substance. Do they know of any alternatives?

Ask customers whether they will accept different substances, as sometimes it is customer demand which dictates processes, etc.

Could different equipment be used which could reduce the harmful effects of the substance?

Could the substance be diluted and still be effective?

Could it be used in a different form, e.g. granules instead of powder, liquid instead of powder, etc.

When an alternative substance has been identified, the following questions may be relevant.

- How effective is the alternative?
- How will it affect the quality of the product?
- Will customers accept it?
- Is the alternative a threat to people's health?
- Is there an immediate threat or is it likely to develop over the long term through frequent exposure?
- Could the new substance increase fire or explosion risks?
- Are you substituting one set of risks for another?
- Is it easier to dispose of the waste?
- How much waste will be generated?
- Will it affect the environment?
- Is the alternative more expensive?
- Will it always be available?
- How volatile is the substance?
- Will the substance create more dust or vapours than the current substance?

Step 3

Look at all of the alternative products identified in Step 2 and ask the question listed for each alternative.

What would be the consequences of using the alternative? Will it do the job as well and as efficiently?

Do you have the equipment to use the new substance or will new sprays, containers, etc. need to be bought? Would this increase the costs and lead to the new substance being a short-term solution?

Will the new substance be available in the long term in the quantities you want? It is best practice to endeavour to keep consistency in substance use, so frequently changing substances could create hazards, as employees will be unfamiliar with what they are using.

What training do you need to give staff on the use of the new substances?

Step 4

Compare the risks of using the alternatives you have identified with the product you are currently using.

Sometimes the hazards of the different products fall into different categories, e.g. one product could be highly flammable and a fire risk and the other could be less flammable but highly toxic.

Try to compare substances in respect of the same hazard, e.g.:

- Is one substance more likely to cause skin disease than another?
- Is one substance more likely to give people cancer than another?

Try to keep comparisons simple:

- Is the substitute going to explode or poison people?
- Will it only affect people who work with it or could it affect other people in the area?
- How likely is it to cause damage to people or the environment?

The general risk assessment for the job activity will identify what residual risks are associated with the job. The substance chosen should not increase those residual risks. It may be necessary to accept a greater risk in one area in order to lessen the risk elsewhere. For example, if working in a critical space is planned and a chemical cleaning agent is needed it will be safer to specify a non-flammable substance because a fire in a confined space is a high risk and the consequences are great (e.g. people often cannot get out easily), but the non-flammable cleaning agent may be toxic. It is better to control the risks of contact with the toxic substance through PPE than to change the toxic substance in favour of a less toxic but highly flammable substance.

Step 5

When all the risks of possible alternatives have been identified and assessed, the decision remains as to whether to substitute the substance or not.

Changing working practices, products and procedures needs careful planning, needs to involve the workforce, may need more training, a change of equipment, etc.

All of these aspects have to be considered in the decision-making process and the decision needs to be taken in view of the 'big picture' — what will be the real, tangible benefits to overall safety of employees and others.

Perhaps it would be wise to test the substitute in localised areas prior to company-wide implementation.

Step 6

If the decision is taken to substitute, then an implementation plan is needed.

A plan should be drawn up which addresses:

- what needs to be done
- who needs to do it
- when it needs to be done
- who needs to know about it.

Organise information, instruction and training for employees.

If the new substance is harmless there will be less risk to employees and the amount of information they will need will be less.

Step 7

Assess the change. All good and effective health and safety management systems have a step of 'Audit and review'.

Is the substance doing what it should do? Is it being used as planned or are employees, for instance, using twice the concentration of the substance to achieve an effective job.

Monitor employees. If the new substance was chosen because it would cut down on skin dermatitis, does it? Can you assess the benefits?

What do the Dangerous Substances and Explosive Atmospheres Regulations 2002 require of employers?

These Regulations (to be known as DSEAR) came into force in December 2002.

They require employers to manage the risks from dangerous substances and explosive atmospheres and apply to all dangerous substances in nearly every building.

The Regulations set minimum requirements for the protection of workers from fire, explosion and risks arising from dangerous substances.

DSEAR complement the requirements of the Management of Health and Safety at Work Regulations 1999 and the general principles of risk assessment in the Management of Health and Safety at Work Regulations 1999 apply to the DSEAR.

Employers and the self-employed must:

- carry out a risk assessment of any work activity involving dangerous substances
- provide technical and organisational measures to eliminate or reduce, as far as reasonably practicable, the identified risks
- provide equipment and procedures to deal with accident and emergencies
- provide information and training to employees
- classify places where explosive atmospheres may occur into zones and mark the zones where necessary.

How does an employer carry out a risk assessment under DSEAR?

DSEAR requires employers (or self-employed persons) to:

- carry out a risk assessment before commencing any new work which involves a dangerous substance
- record the findings in writing if there are five or more employees as soon as is practicable after the assessment

- take steps to eliminate the risks
- ensure that the workplace and work equipment is safe during operation and maintenance
- detail any hazardous zones
- detail any special measures of co-operation between more than one employer
- introduce arrangements for dealing with accidents and emergencies.

The risk assessment is an identification and careful assessment and examination of the dangerous substances present in the workplace, the work practices and activities using those substances and what the consequences are if something goes wrong.

The risk assessment must consider the risks not only to employees but also to members of the public.

Employers need to establish what they need to do to reduce or eliminate risks as far as is reasonable practicable so as to ensure everyone's safety when using dangerous substances.

The risk assessment *must* be completed before work commences with a dangerous substance and the control measures identified as being necessary must be implemented *before* works commence.

The risk assessment process should be the same as for all risk assessments:

- identify the hazard
- identify who may be harmed and how
- evaluate the risks and decide on the control measures necessary
- record the findings
- audit and review the risk assessment.

What are some control measures which could be implemented?

The following list gives some examples of control measures but the risk assessment should help to identify what controls are needed in the actual environment in which the dangerous substance is present.

- Reduce the quantity of dangerous substances to a minimum.
- Avoid or minimise releases.
- Control releases at source.
- Prevent the formation of an explosive atmosphere.
- Collect, contain and remove any releases to a safe place.
- Avoid ignition sources.
- Avoid adverse conditions that could lead to danger, e.g. avoid excessive temperatures.
- Keep incompatible substances apart.

Other steps which could be taken will mitigate the risks associated with a dangerous substance.

- Reduce the number of employees exposed.
- Provide plant that is explosion-proof.
- Provide explosion suppression or explosion relief equipment.
- Take measures to minimise or control the spread of fires or explosions.
- Provide suitable personal protective equipment.
- Design and construct the workplace to reduce the risks from dangerous substances.
- Choose appropriate equipment and work systems which reduce risks.
- Implement safe systems of work, e.g. Permit to Work procedures.

How will an employer identify a dangerous substance?

You will need to carry out two steps.

(1) Check whether the substance has been classified as
  - ○ explosive
  - ○ oxidising

- extremely flammable
- highly flammable or
- flammable.

(2) Assess the physical and chemical properties of the substance and the circumstances of the work involving the substance, to see what will create a safety risk to persons.

Any substance labelled:

- explosive
- oxidising or
- flammable (in all categories)

is a dangerous substance and DSEAR apply.

CONTROL OF SUBSTANCES HAZARDOUS TO HEALTH ASSESSMENT SHEET

Company:

Address:

Contact:

Product:

Job task:

Application:

Equipment:

CONTROL MEASURES

For users:

For persons in location:

EMERGENCY PROCEDURE

Emergency contact number:

Spillage arrangements:

Consumption arrangements:

Contact arrangements:

PRODUCT:

SAFETY DATA SHEET ATTACHED: YES/NO

RISK IDENTIFICATION

Hazardous component(s):

Hazardous nature of component(s):

Health hazards (known):

Persons affected:

Duration of exposure:

Level of exposure:

Risk category:

OCCUPATIONAL EXPOSURE LIMITS

Occupational Exposure Standard:

Maximum Exposure Limit (8-hour TWA):

Maximum Exposure Limit (15-minute TWA):

CONTROL MEASURES OF EXPOSURE:

MONITORING REQUIREMENTS:

OTHER CONTROL MEASURES

Training:

Health surveillance:

Re-assessment:

Date of assessment/revision:

6

Manual handling

What are employers responsible for with respect to manual handling at work?

The Manual Handling Operations Regulations 1992 apply to all manual handling activities carried out by employees while at work.

Employers must, as far as is reasonably practicable, avoid the need for employees to undertake any manual handling activities while at work which involve risk of injury.

Despite the above requirement, manual handling is a common cause of work-related injury. In some cases, poor manual handling can lead to permanent disability and physical impairment.

Employers must undertake a risk assessment of all manual handling activities and determine a hierarchy of risk control in order to minimise injury and ill-health risks.

Information on the weight of objects to be manually handled must be given to employees. This can be general information or more specific or precise product information. Many manufacturers and suppliers of products and equipment are displaying the weight of the item on packaging or on delivery notes, etc.

What are the costs of poor manual handling to both businesses and society?

Injuries sustained by employees while they are handling, lifting or carrying items at work account for 38% of all notified 'over three day' injuries.

Over 1 million people are reported to have suffered 'illness' from musculo-skeletal disorders and the prevalence rate is increasingly when compared with the early 1990s.

Statistics from the HSE for 2001/2002 show that approximately 12.3 million days were lost in employment productivity due to musculo-skeletal disorders.

The National Health Service has one of the highest incidence rates for musculo-skeletal injury and approximately 50% of all reported 'over three day' injuries were due to injury while handling, lifting and carrying, i.e. approximately 5000.

Back injury is not the only type of injury to be sustained from manual handling. Injuries are reported which affect:

- hands
- feet
- arms
- legs
- neck
- head.

Many manual handling injuries are the result of poor practices being followed over lengthy periods of time and not from a 'one-off' manual handling activity.

On average, each injury takes 20 days for recovery and, in some instances, disability is permanent. Costs to business will be huge and are often hidden in real terms. Costs to be considered are:

- costs of sick pay
- cost of loss of skilled employee
- replacement/temporary staff
- reduced productivity

Case study

An ambulance worker received compensation in 2002 of £140 000 in an out-of-court settlement with his employers for serious back injuries sustained in the course of his employment.

The employee was lifting a patient when two wheels came off the stretcher he was carrying. He then had to bear the patient's weight for five minutes.

As a result, he damaged his lower back and right leg.

Damages were claimed against the NHS Trust because the stretcher had been modified to fit into the ambulance and was not fit for its purpose. The Trust admitted liability.

Consideration is being given by the NHS Trust to instigating legal action against the stretcher manufacturer.

- investigation time
- civil claims
- criminal prosecution
- increased insurance premiums.

Back injuries represent the biggest single group of claims for incapacity benefit.

Costs for manual handling injuries have been estimated at £6 billion in lost production.

What steps should be taken in respect of carrying out a risk assessment for manual handling?

In the first instance, it would be sensible to conduct a general assessment to see if manual handling activities give rise to hazard and risk, as not *all* manual handling will.

Remember that manual handling includes:

- lifting
- pushing
- pulling
- shoving
- lowering
- carrying.

Consider the size and shape of the load and the best way to handle it:

- If the load is difficult or heavy, seek assistance.
- Consider where the load is going — is the pathway clear and free from obstruction.
- Is the place ready where the load is to go?
- Can lifting devices be used?
- Can the load be split to make carrying easier?
- Manual handling involves pushing and pulling as well as lifting. Can any of these jobs be mechanised?
- Complete a manual handling risk assessment.

What needs to be considered for a detailed manual handling risk assessment?

The tasks

Do the tasks involve:

- holding loads away from the body
- twisting, reaching or stooping
- strenuous pushing or pulling
- unpredictable movement of loads
- large vertical movement
- long carrying distances

- repetitive handling
- insufficient rest time
- a work rate imposed by a process?

The loads

Are the loads:

- heavy, bulky or unwieldy
- difficult to hold
- unstable or unpredictable
- intrinsically harmful, e.g. sharp?

The working environment

In the working environment, are there:

- constraints on posture
- variations in level
- poor floors
- hot, cold or humid conditions
- strong air movements
- poor lighting
- restrictions on movement or posture from clothes or personal protective equipment?

Individual capacity

Does the job:

- require unusual capability
- endanger those with a health problem
- endanger pregnant women
- require special information or training?

Case study

Types of manual handling in licensed premises

The following activities undertaken routinely in most pubs are likely to present particular risks from manual handling operations:

- the delivery and removal of full and empty kegs, boxes, barrels, crates and gas cylinders
- the stacking of full kegs and barrels
- the movement of kegs, barrels, etc. within the cellar or storeroom
- shifting of casks
- the movement of loads between floors — carrying crates from the cellar to the bar
- lifting buckets of water or pipe cleaning containers
- lifting gas cylinders
- putting items on shelves and getting items off shelves
- movement of furniture, equipment, etc.
- food deliveries
- removal of glass bottle skips
- carrying empty glass baskets or crates
- changing optics
- lifting glass washer trays
- carrying tills or money drawers
- carrying money or change
- moving gaming machines
- assisting with entertainment equipment.

What are some of the ways of reducing the risks of injury from manual handling?

Some ways of reducing the risk of injury are detailed below.

The tasks

Can you:

- reduce the amount of twisting and stooping
- avoid lifting from floor level or above shoulder height
- avoid strenuous pushing or pulling
- reduce carrying distances
- avoid repetitive handling
- vary work, allowing one set of muscles to rest while another is used?

The loads

Can you make the loads:

- lighter or less bulky
- easier to hold
- more stable
- less damaging to hold
- have you asked your suppliers to help?

The working environment

Can you:

- improve workplace layout to improve efficiency
- remove obstructions to free movement
- provide better flooring

- avoid steps and steep ramps
- prevent extremes of hot and cold
- improve lighting
- consider less restrictive clothing or personal protective equipment?

Individual capacity

Can you:

- take better care of those with physical weaknesses or who are pregnant
- give your employees more information, e.g. about the range of tasks they are likely to face
- provide training?

What are good handling techniques?

The following are important points to bear in mind when handling a load, using a basic lifting operation as an example.

Planning

Plan the lift. Where is the load to be placed? Use appropriate handling aids if possible. Do you need help with the load? Remove obstructions. For a long lift, such as floor to shoulder height, consider resting the load midway on a table or bench to change grip.

Positioning feet

Feet should be apart, giving a balanced and stable base for lifting (tight skirts and unsuitable footwear make this difficult). The leading leg should be as far forward as is comfortable and, if possible, should be pointing in the direction you intend to go.

Good posture

When lifting from a low level, bend with the knees, but do not kneel or over-flex the knees. Keep the back straight, maintaining its natural curve (tucking in the chin helps). Lean forward a little over the load if necessary to get a good grip. Keep the shoulders level and facing in the same direction as the hips.

Holding the load

Try to keep the arms within the boundary formed by the legs. The best position and type of grip depends on the circumstances and individual preference, but must be secure. A hook grip is less tiring than keeping the fingers straight. If you need to vary the grip as the lift proceeds, do it as smoothly as possible.

Lifting

Keep the load close to the trunk for as long as possible. Keep the heaviest side of the load next to the trunk. If a close approach to the load is not possible, slide it towards you before you try to lift. Lift smoothly, raising the chin as the lift begins, keeping control of the load.

Movement

Move the feet instead of twisting the trunk when turning to the side.

Adjustment

If precise positioning of the load is necessary, put it down first, then slide it into the desired position.

Risk assessment: 20 litre drum handling

The task

Between twenty and thirty 20 litre drums were moved from the delivery area to the storage area. They were then moved from the storage area to the point of use.

The problem

To move the 20 litre drums to the storage area, maintenance staff had to carry them down 20 steps. No injuries had been reported but the task was identified as having a high potential risk of musculo-skeletal injuries.

The solution

The storage area was reorganised and moved to be near the delivery area so that the need to carry the drums down the steps was eliminated.

Two sack barrows were provided so that the drums could be easily moved on the same level to the areas where they are needed.

The benefits

Easier manual handling.

More efficient use of time — less double handling.

Reduction in potential manual handling injuries.

What are the guideline weights for lifting or manual handling?

	Female:	Male:
(1) Shoulder height		
— arms extended	3 kg	5 kg
— near to body	7 kg	10 kg
(2) Elbow height		
— arms extended	7 kg	10 kg
— near to body	13 kg	20 kg
(3) Thigh height		
— away from body	10 kg	15 kg
— near to body	16 kg	25 kg
(4) Knee height		
— away from body	7 kg	10 kg
— near to body	13 kg	20 kg
(5) Lower leg height		
— away from body	3 kg	5 kg
— near to body	7 kg	10 kg

Each category indicates the guideline weights for lifting and lowering loads.

Heavier weights can be handled more safely if they are held close to the body. Carrying objects at arm's length creates extra strain of the spine and muscles and therefore lower weights are recommended.

The weights assume that the load is readily grasped with both hands and that the operation takes place in reasonable working conditions with the lifter in a stable body position.

Any operation involving more than twice the guideline weights should be rigorously assessed — even for fit, well-trained individuals working under favourable conditions.

Twisting

Reduce the guideline weights if the lifter twists to the side during the operation. As a rough guide, reduce them by 10% if the handler twists beyond 45°, and by 20% if the handler twists beyond 90°.

Frequent lifting and lowering

The guideline weights are for infrequent operations — up to about 30 operations per hour — where the pace of work is not forced, adequate pauses to rest or use different muscles are possible, and the load is not supported for any length of time. Reduce the weights if the operation is repeated more often. As a rough guide, reduce the weights by 30% if the operation is repeated five to eight times a minute; and by 80% where the operation is repeated more than twelve times a minute.

Exceeding the guidelines

The risk assessment guidelines are not absolute safe limits for lifting. But work outside the guidelines is likely to increase the risk of injury, so you should examine it closely for possible improvements. You should remember that you must make the work less demanding if it is reasonably practicable to do so.

Is 'ergonomics' anything to do with manual handling?

Yes and no! Ergonomics is the science concerned with the 'fit' between people and their work and surroundings.

Ergonomics aims to make sure that tasks, equipment, information and the environment suit each worker. So, that could include manual handling activities but it is more likely to consider *how* the job is done rather than *what* is lifted.

How can ergonomics improve health and safety?

Applying ergonomic principles to the workplace will:

- reduce the potential for accidents
- reduce the potential for injury and ill-health
- improve performance and productivity.

Equipment, controls, operating panels, isolation switches, etc. should all be designed for ease of use but, in practice, how many times are switches awkward to get at, requiring twisting and contortion to use them?

A machine with a control panel which the operator is required to use could be the cause of accidents and injury if:

- the switches and buttons could be easily knocked on or off, thereby starting or stopping the machine by mistake
- the warning lights or switches are unusual colours or the opposite colours to those usually expected, e.g. if red is for 'go', green is for 'danger'. Also, colours may be important as many people have colour blindness for red and green
- the instruction panel and information given on how to use the controls is complicated or too detailed, causing operator confusion and inappropriate actions.

Ergonomics would look at all of the above issues and 'design out' the hazards associated with the control panel on the machine. The location of controls would be considered so as to cut down on 'repetitive strain' injuries.

What kind of manual handling problems can ergonomics solve?

Ergonomics is typically thought of as solving physical problems and in respect of manual handling these problems would be:

- loads which are too heavy or bulky
- loads which need to be lifted from the floor or from above shoulder height

- repetitive lifting
- tasks which involve awkward postures, twisting or bending
- loads which cannot be gripped properly
- tasks which need to be carried out in poor environmental conditions, i.e. wet floors, poor lighting, cramped space, restricted headroom
- tasks carried out under too great time pressures and without adequate rest periods.

Any of the above situations can lead to operator tiredness and exhaustion. This increases the risk of accidents and injury.

Ergonomics is about finding solutions for alternative ways of doing the job.

Is it appropriate to have a 'no lifting' policy within the work environment?

'No lifting' policies seem to be common in the care services industry where there is a lot of manual handling of people and equipment.

Such an all-encompassing policy may be workable in some organisations but it may be extremely difficult to enforce such a policy in reality.

The Manual Handling Operations Regulations 1992 require that 'hazardous' lifting or manual handling is eliminated or reduced to an acceptable level. Manual handling in all its forms cannot be eliminated, but controls can be put in place to reduce the likelihood of injury.

A 'no lifting' policy would require mechanical aids to be provided to assist with lifting. In some circumstances the use of a lifting device may create a greater hazard that the lift itself.

Rather than a 'no lifting' policy it would be appropriate to have a 'lifting' policy which sets out what type of manual handling is undertaken, what the hazards and risks are and how control measures can be used to reduce the risks.

Case study

Risk assessment format for care services

Consider activities to be undertaken during the day and also at night.

A risk assessment for use in the care services (e.g. for a care home) should be set out in a simple format so that it is possible to quickly assimilate what equipment, techniques and numbers of staff are appropriate.

The following should be included:

- individual details including height and weight
- the extent of the individual's ability to support his or her own weight and other relevant factors, e.g. pain, disability, tendency to fall, etc.
- problems with comprehension or co-operational behaviour
- recommended methods for relevant tasks such as sitting, visiting the toilet, bathing, transfers, movement in bed
- the minimum number of staff needed to help
- the need for and availability of lifting or moving equipment
- other relevant risk factors
- what level of training and individual capability is required of the care worker.

What are some of the key solutions to manual handling problems?

There are several ways in which manual handling problems can be reduced or eliminated.

These include:

- avoiding manual handling through automation or changing the overall process
- re-designing the load
- re-designing the task
- re-designing the working environment
- introducing mechanical handling aids.

The changes do not have to be expensive or complicated to be effective. Simple solutions are often best.

Principles for developing successful solutions to manual handling problems

- Prioritise your activities.
- Tackle serious risks affecting a number of employees before isolated complaints of minor discomfort.
- Find a few possible solutions to evaluate.
- Try out ideas on a small scale and modify them if necessary prior to full implementation.
- Monitor the solutions to make sure they remain effective.
- Keep abreast of new technologies.

MANUAL HANDLING RISK ASSESSMENT FORM

Job description:

Who is undertaking the tasks?

What are the hazards involved in the job?

What are the risks?

How likely are the risks?

What control measures are needed to reduce or eliminate the risks?

Those currently in place:

Those which need to be implemented:

When should control measures be implemented?

When should the risk assessment be reviewed?

Date assessment completed:

By whom:

7

Noise

What are the requirements of the Noise at Work Regulations 1989?

Sounds and noise are an important part of everyday life. In moderation they are harmless but if they are too loud they can permanently damage your hearing. The danger depends on how loud the noise is and how long you are exposed to it. The damage builds up gradually and you may not notice changes from one day to another, but once the damage is done, there is no cure. The effects may include:

- sounds and speech may become muffled so that it is hard to tell similar sounding words apart, or to pick out a voice in a crowd
- permanent ringing in the ears (called tinnitus)
- a distorted sense of loudness — sufferers may ask people to speak up then complain that that they are shouting
- needing to turn up the television too loud, or finding it hard to use the telephone.

The law requires employers to safeguard the health, safety and welfare of their employees while they are at work. The employer must provide a working environment which is safe and without risks to health.

So, any environment which is subject to excessive noise will be unsafe because there is a risk of noise-induced hearing loss.

The Noise at Work Regulations 1989 have set down more specific requirements on controlling noise and employers must carry out risk assessments, eliminate noise at source or reduce it to tolerable and 'safe' levels.

Failure to protect employees' hearing is an offence and carries fines in both the magistrates' and Crown Courts.

What are the hazards from noise?

Hearing damage

Exposure to high noise levels can cause incurable hearing damage. Usually, the important factors are:

- the noise level, given in decibel units as dB(A)
- the length of time over which people are exposed to the noise: daily and over a number of years.

Sometimes the peak pressure of the sound wave may be so great that there is a risk of instantaneous damage. This is most likely when explosive sources are involved, such as in cartridge-operated tools or guns.

The damage involves loss of hearing ability, possibly made worse by permanent tinnitus and other effects. Sufferers find it hard to distinguish words clearly, e.g. they tend to confuse words such as 'bit' and 'tip'.

Other effects of noise at work

Noise at work can cause other problems, such as disturbance, interference with communications and stress. Although the Noise at Work Regulations 1989 do not deal with these specifically, you should bear in mind that they might also need to be tackled.

What do employers have to do about noise?

Noise must be assessed to establish whether employees are subjected to unacceptable levels which could cause permanent hearing damage.

The persons carrying out noise assessments must be competent to do so. Environmental assessments may be appropriate but so might personal dose meters to establish exactly what each individual is exposed to.

Noise of all descriptions must be assessed, as must the cumulative effects of noise from numerous sources.

Is there anything I must know before undertaking a noise assessment?

Yes. An employer must be fully conversant with the Noise at Work Regulations and must understand the different noise levels at which action must be taken. These are called *action levels*.

Employers have a duty to reduce noise levels to the lowest practicable level.

Noise levels are calculated on the sound emitted from machines, processes, etc. and the length of time employees are exposed. Individuals can be exposed to high sound levels for short periods of time and not suffer hearing damage. The longer the exposure time to noise, the greater the risk of hearing damage.

Noise levels are measured in decibels (dB(A)). If the noise level in the workplace is below 85 dB(A) then the employer does not have to do anything specific to control the noise, merely be sure that the noise is as low as it can get.

If noise levels are above 85 dB(A) but below 90 dB(A), the employer has to carry out a noise risk assessment. Assessments must be done by competent persons. At this level, employees may request hearing defenders and the employer must provide them.

If noise is above 90 dB(A), a hearing protection zone must be declared and hearing defenders provided to all employees and visitors. Suitable safety signs must be displayed in hearing protection zones.

More stringent requirements are imposed on higher noise levels.

Are these noise action levels the permanent levels or will they change?

A new EU Noise Directive was tabled at various European Council meetings during 2000/2001 and, in November 2001, a new directive on noise was adopted by member states. This directive will repeal the earlier directive on which the Noise at Work Regulations 1989 were based. The UK will need to introduce new Noise at Work Regulations and this will have to be achieved by 2006.

The proposed Regulations will introduce new 'action levels' as follows:

(1) provide information and training to workers at 80 dB(A) (currently 85 dB(A))
(2) workers will have the right to hearing checks/audiometric testing at 85 dB(A) (as now) and also at 80 dB(A) as the risk is indicated
(3) make hearing protection available at 80 dB(A) (currently 85 dB(A))
(4) hearing protection to be worn at 85 dB(A) (currently 90 dB(A))
(5) limit on exposure to noise to be 87 dB(A) (currently no limit)
(6) programme of control measures at 85 dB(A) (currently 90 dB(A))
(7) designate noise control areas, display notices, etc. at 85 dB(A) (currently 90 dB(A))
(8) noise exposure which varies daily can be averaged over a week (currently eight hours).

Action levels are therefore going to reduce. This may seem achievable — only a 5 dB(A) reduction. But, in noise level terms, a 3 dB(A) rise in noise level is, in effect, a doubling of the sound pressure levels. In simple terms, a lowering of noise action levels to 80 and 85 dB(A) will be quite significant.

How should a noise assessment be carried out?

Decide whether you might have a problem

If people have to shout or have difficulty being understood by someone about two metres away, you might have a problem. To be sure about this you will need to get the noise assessed.

Get the noise assessed

Your assessment should find out whether noise exposure is likely to reach the action levels. Where the assessment shows you have a noise problem, you should use it to help you develop plans for controlling exposure.

The job must be done by a competent person, someone who understands the Health and Safety Executive's guidance on assessment and how to apply it in the workplace. The essential qualification for the person is the ability to do the job properly and to know his or her own limits; this is more important than formal qualifications. However, many technicians may need extra training and local technical colleges often provide short courses lasting a few days or can advise on where they are available. Alternatively you might call a consultant.

Tell the workers affected

Where your assessment shows exposure is at or above any of the action levels, you should let employees know there is a noise hazard and what you want them to do to keep risks to a minimum.

Reduce the noise as far as reasonable practicable

Where the exposure needs to be controlled, the most reliable way is to quieten the workplace if this can be done.

You can avoid problems if you can make sure that noise reduction is built into new machinery when you buy it. Ask about noise before deciding which machine to buy.

You should also consider whether it might be possible to reduce either the number of people working in noisy areas or the time they have to spend in the areas. Perhaps some jobs can be done in a quieter location.

Ear protection

If people have to work in noise-hazardous areas, they will need ear protectors (ear muffs or ear plugs). However, these should not be regarded as a substitute for noise reduction. As long as people work in noise at or above the second or the peak action level, the Regulations still require you to reduce the noise exposure by other means as far as this is reasonably practicable.

Between the first and second action levels, you should make sure that:

- protection is freely available
- the employees know that unless they wear it, there is some risk to their hearing.

The Regulations do not, however, make it a legal duty for employees to wear protection below the second action level.

Make sure that young people in particular get into the routine of wearing ear protectors before their hearing is damaged.

Where use of protection is compulsory, ear protection zones should be marked if this is reasonably practicable. Ensure that everyone who goes into a marked zone, even for a short time, uses ear protection.

Check to make sure your programme is working

Make sure the equipment you provide is kept in good condition.

If you rely on ear protectors, find out whether they are really being used. If anything is wrong do not neglect it, put it right!

What information should be given to employees?

Employees must, by law, be provided with information about any risk to their hearing, especially if they will be exposed to levels of 85 dB(A) or above.

Adequate information, instruction and training is required so that the employee understands:

- the risk of damage to their hearing
- the steps which they can take to minimise the risk
- the procedure they need to follow to obtain personal ear protection
- their own duties under the Regulations.

The five steps to risk assessment

- Identify the hazards.
- Identify the people affected.
- Evaluate the risks and identify the control measures necessary.
- Record findings in writing.
- Audit and review.

What types of ear protection are available?

There are generally two main styles or types of ear defender:

- the 'headphone' type which cover the ears completely
- ear plugs which fit into the ear canal.

Ear protectors will only be effective if they are in good condition and properly maintained.

They must suit the individual and be worn properly. It is vitally important therefore to consult employees on what types they prefer — it should be an individual choice.

Do employers have other responsibilities besides providing ear protection?

Yes. The main responsibility of employers is to reduce the noise *at source* to the lowest level possible.

Noise can be controlled by:

Engineering controls	Purchasing equipment with low noise emissions. Changing the process, e.g. presses instead of hammers. Avoiding metal to metal impacts. Using flexible couplings and mountings. Introducing design dampers. Correct sizing of ductwork, fans, motors, etc.
Orientation and location	Move the noise source away from employees, turning machines around so that noise or sound waves can travel out of the building.

	Not putting machines, etc. into hard surface areas as noise 'bounces' off surfaces.
Enclosure	Surround the machine or noise source in sound-absorbing material (total enclosure is most effective).
	Soundproof the room/work area.
	Introduce sound-absorbent materials to surfaces.
Use of silencers	Use on ductwork, for motors.
	Use on pipes which carry gas, air or steam.
	Use on exhaust ventilation systems.
Lagging	Lag pipes as an alternative to enclosure.
Damping	Dampers can be fitted to ductwork.
	Use double skin design, preferably with noise-absorbent material in between.
Absorption	Acoustic ceiling and wall panels help to absorb the sound waves.
Screens	Temporary acoustic screens can help to reduce levels of noise and these can be moved to where needed.
Isolate workers	Remove workers from the noise source by construction acoustic booths for them to work in.

Usually it will need specialist noise or acoustic consultants to work out exactly what needs to be done to reduce noise to safe levels. Remember, noise must be a combination of different machines.

If you feel that you have 'a din' in the workplace, then you will need expert advice to reduce noise levels to tolerable levels.

Who can carry out a noise assessment?

Regulation 4 of the Noise at Work Regulations 1989 requires that an employer appoints a competent person to carry out noise assessments.

A competent person needs to have:

- knowledge
- experience
- information
- inter-personal skills.

In particular, the competent person should show skills appropriate to the situations to be handled, including:

- an understanding of the purpose of assessments
- a good basic understanding of what information needs to be obtained
- an appreciation of their own limitations, whether knowledge, experience, facilities or resources
- how to make measurements
- how to record results, analyse and explain them to others
- the reasons for using various types of measuring instruments and their benefits
- how to maintain, check and attenuate equipment
- how to interpret information provided by others and how to assimilate it into other data so as to give overall results of noise levels, etc.

The competent person will *not* need an advanced knowledge of acoustics.

Nor will they necessarily need to know the full details of how to guide and advise the employer on where to obtain further specialist advice.

Competence, as always, is judged in relation to the complexities of the situation to be assessed and should not be over-exaggerated.

Assessing noise hazards in industrial production plants which involve many different processes will be more difficult than on a construction site where a few power tools are in use.

Can someone from within the workforce be appointed as the competent person?

Yes, if they possess the necessary knowledge, skills, experience and information to be able to assess the hazards, risks and control measures necessary in respect of noise at work.

Formal qualifications in acoustics are not necessary except for the most complex of work environments.

A commonsense approach to identifying hazards and assessing risks is often more practical than in depth subject knowledge.

An employee may need to receive further training in risk assessment techniques, the use of sound level meters, personal dose meters, etc.

Are there any typical topics which an employee or other person needs to be trained in so that they can carry out noise assessments?

Some typical topics for a training course are given below.

Other sources of information.

Legal requirements	Information on Noise at Work Regulations 1989, Management of Health and Safety at Work Regulations 1999, the Health and Safety at Work Etc. Act 1974.
Purpose of noise assessment	Need to know how to assess exposure to noise.

Requirement for noise assessment	Information on sound, noise, sound pressure waves, hazards, risks, etc. Different types of noise exposure, e.g. daily dose levels, dB(A) scales, etc.
Equipment for measuring noise	Sound level meters, personal dose meters.
Noise measuring procedures	Survey techniques, survey procedures, sampling process, location of microphones, etc., measurement of peak noise levels.
Calculation of noise exposure	Calculation of noise levels over 8-hour period, peak action levels.
Noise sources and control measures	How to control noise, risk reduction, etc.
Ear protection	Types and use of ear protection.

What is noise exposure?

Any audible sound should be considered as noise and be included in the assessment of noise exposure. This includes speech, music, noise from communications devices, noise from machinery, background processes, traffic, etc.

The 'daily personal noise exposure' is a measure of noise energy received by a person during the working day.

Noise exposure must also address any 'peak' sounds which happen infrequently during the day, e.g. impact noise, explosive noise, etc.

Who needs to be assessed for noise exposure?

All employees who may or will be exposed to noise levels during the course of the working day which exceed 85 dB(A) must be formally assessed.

If the noise in the workplace is so loud that, generally, employees have to shout to one another to be heard, the chances are that the noise levels will be above 85 dB(A) and a formal Noise Assessment will be needed.

As with all risk assessments, the effect of the job task or activity on persons other than employees also needs to be assessed.

What actually are decibels?

The ear can hear sounds at frequencies between 20 and 20 000 cycles per second or hertz (Hz). It is most sensitive to frequencies between 3000 and 6000 Hz, i.e. those used in human speech.

Sound levels are measured in decibels (dB) with the range going from zero decibels (the threshold of hearing) up to around 140 dB (a very painful and dangerous level of exposure).

A correction is made to allow for the human ear's varying ability to hear sounds at different frequencies. This is called the 'A' weighting and noise levels corrected in this way are shown as dB(A).

The dB(A) scale is therefore noise as it is heard by the human ear.

The decibel scale is logarithmic. A rise of 10 dB(A) represents a tenfold increase in noise.

An increase of 3 dB(A) results, in effect, in the doubling of the noise. A small rise in dB(A) can significantly increase the hazards and risks from the noise source.

Typical sound levels: dB(A)

0	Faintest, audible sound
10	Leaf rustling, quiet whisper
20	Very quiet room, e.g. library
30	Subdued speech
40	Quiet office
50	Normal conversation
60	Busy office
70	Loud radio or TV
80	Busy street in daytime
90	Heavy vehicle close by
100	Bandsaw cutting metal
110	Woodworking, industrial machine shop
120	Chainsaw
130	Riveting
140	Jet aircraft taking off close by

NOISE RISK ASSESSMENT FORM

Company details . . . .

Business activity . . . .

Number of employees . . . .

Person responsible for noise assessments or competent person . . . .

Area being assessed . . . .

Describe work activity . . . .

Number of employees . . . .

Number, type and use of machines . . . .

Name of person carrying out assessment . . . .

Date of assessment . . . .

Equipment used for assessment . . . .

Calibration details . . . .

Model and reference number . . . .

Location of noise assessment equipment . . . .

Duration of survey/assessment . . . .

Exposure assessment measurements . . . .

Describe machine types	**Noise levels: dB(A)**

Maximum exposure times (mins/hours) . . . .

Has the first action level been exceeded, i.e. 85 dB(A)? Yes/No

Has the second action level been exceeded, i.e. 90 dB(A)? Yes/No

What control measures are in place?. . . . .

. . . .

What control measures are needed in addition to those already in place? . . . .

. . . .

. . . .

To what level will noise be reduced once control measures are in place? . . . .

. . . .

. . . .

When should remedial measures be completed? ....

....

When should noise assessments be reviewed? ....

....

Other comments, e.g. is health surveillance necessary, etc.? ....

....

....

....

Date noise assessment completed ....

Name of assessor ....

Contact details ....

....

8

Display screen equipment

What are the requirements of the Health and Safety (Display Screen Equipment) Regulations 1992?

The above Regulations require employers to minimise the risk arising from working with display screen equipment by ensuring that workplaces and jobs are well designed and that equipment is suitable and sufficient and chosen so as not to cause the risk of injury or ill health.

The Regulations require every employer to carry out a risk assessment of display screen equipment so that hazards and risks can be identified and control measures implemented.

Workstations have to meet minimum requirements.

Work patterns have to be adapted so that display screen equipment users can have regular breaks away from their screens.

Employers must provide eyesight tests to those employees or users who require them and must provide corrective spectacles where they are needed for the display screen equipment use.

All employees and other users (e.g. operators) must be given suitable health and safety training and information.

What are the health problems associated with using display screen equipment?

Health problems are not always immediately obvious when using display screen equipment as the symptoms may be quite minor at first but the repetitive nature of the tasks can exacerbate minor injuries until they become quite debilitating.

Health problems associated with using display screens are:

- upper limb disorders, including pains in the neck, elbows, arms, wrists, hands and fingers
- backache
- headaches and migraines
- eye strain but *not* eye damage
- fatigue and stress.

Aches and pains in limbs are often referred to as repetitive strain injuries (RSI) and include carpel tunnel syndrome and tennis elbow.

Do the Regulations affect everybody who uses display screen equipment?

No. The Regulations in the main apply to the *users* of display screen equipment. The Regulations also only apply to employees and the self-employed. Therefore a 'user' can only be an employee or a self-employed person.

Regulation 1(2d) defines a 'user' as an employee who habitually uses display screen equipment as a significant part of his/her normal work.

In the same Regulation, a self-employed person who habitually uses display screen equipment is defined as an 'operator'.

A display screen is not only a computer screen but also television screens, video screens, plasma screens, microfiche screens, etc.

Emergency technologies are creating new types of screens and it is anticipated that the Regulations will cover all of these.

Employers must decide who is a user or an operator under the Regulations and apply the requirements of the relevant Regulations.

Workers who do not input or extract information from a display screen are generally not users.

Employers need to ask themselves a few searching questions in order to ascertain whether they have users or operators on their staff.

First question:	Do any of my employees or the people whom I engage as 'contractors' normally use display screen equipment (DSE) for continuous or near-continuous spells of an hour or more at a time?
Second question:	Do any of them use DSE for an hour or more, more or less daily?
Third question:	Do they have to transfer information quickly to or from the DSE?
Fourth question:	Do they need to apply high levels of attention and concentration to the work that they do?
Fifth question:	Are they highly dependent on their DSE to do their job or have they little choice in using it?
Sixth question:	Do they need special training or skills to use the DSE?

Part-time or flexible workers must be assessed on the same criteria because it is not the length of time they spend 'at work' which counts but the length of time they spend using the VDU or DSE.

Sometimes, employers may wish to simplify things and class *all* users of display screen equipment as 'users' or 'operators' under the terms of the Regulations. This means that the good practice requirements of the Regulation will be applied throughout the organisation.

Examples of display screen users

- Typist, secretary, administration assistant who uses a PC or word processor for typing documents, etc.
- Word processing worker
- Data entry clerk/operator
- Database operator and/or creator
- Telesales personnel
- Customer service personnel if computer entry of information is a common part of the job
- Journalists, editorial writers
- TV/video editing technicians
- Micro-electronics testing operators who use DSE to view test results, etc.
- CAD technicians
- Air traffic controllers
- Graphic artists
- Financial dealers

Is it easy to define who is not a user of display screen equipment?

Yes, relatively so.

The answers to the six questions listed previously should enable the employer to easily differentiate who is who.

Anyone who uses display screen equipment occasionally will not be a user under the Regulations. Nor will anyone who can choose when or for how long they use DSE.

Laptop users will probably *not* be users as they can (usually) choose when, where and for how long they use their screens and computer.

Receptionists will often not be classed as users as they are not continuously using their screens (unless they predominately operate a switchboard which relies on a screen for extension transfers, etc.).

Are employees who work at home covered by the Regulations?

If they are an employee and they use their display screen equipment continuously as part of their job, they will be defined as a user irrespective of where they use the equipment.

The display screen and workstation does not have to be supplied by the employer — an employee can provide their own equipment but the employer would still have to comply with their duties in respect of users and operators and assess the hazards and risks to health.

In order to determine whether homeworkers are users or operators of display screen equipment the six questions posed earlier will need to be asked of each individual worker.

What is a 'workstation' and how do the Regulations apply to these?

A workstation is defined in Regulation 1 of the Health and Safety (Display Screen Equipment) Regulations 1992 as:

> an assembly comprising:
> (i) display screen equipment
> (ii) keyboard or other input device
> (iii) optional software
> (iv) optional accessories to the display screen equipment
> (v) any disk drive, telephone, modem, printer, document holder, work chair, work desk, work surface or other item peripheral to the display screen equipment
> (vi) the immediate work environment around the display screen equipment.

Regulation 2 requires employers to perform a suitable and sufficient analysis of workstations which:

- are used for the purposes of this, his undertaking (regardless of who provided them), by users
- have been provided by him and are used by operators

in order to assess the health and safety risks to which those people are exposed as a result of that use.

The analysis is to assess and reduce risk — it is a risk assessment.

Is it necessary to complete a risk assessment for each workstation and user or operator?

Yes, because each person is different and the effect that using the display screen *may* have on them will be different for individuals. Of course, some individuals may have no ill-effects from using a display screen and there will be little you will need to do.

Individual workstations may vary in design, people's tasks will be different, the amount of control they have over their jobs may be different.

The most effective way to conduct risk assessments for display screen equipment users is to create a questionnaire which includes all the relevant sections on:

- display screens
- keyboards
- mouse or trackball
- software
- furniture
- environment.

The workstation analysis or risk assessment is best done by the individual concerned, once they have had proper training in what they are to look for and how to record the information.

Do the Regulations only require an employer to carry out these risk assessments?

No, they are one part of the employer's responsibilities under the Regulations.

The Regulations themselves require employers to:

- analyse workstations to assess and reduce risks
- ensure that workstations meet minimum specified requirements
- plan work activities so that they include short breaks or changes of activity
- provide eye and eyesight tests on request and special spectacles if needed
- provide information and training.

What are the 'minimum specified requirements' for workstations?

The Regulations are quite specific about requirements for workstations and the appropriate Regulation, Regulation 3, was amended in 2002 to address an European Ruling on the interpretation of the Regulation applying to workstations.

The 'minimum specified requirements' apply to all workstations provided by an employer and not just to those used by 'users or operators'.

The European Court, in effect, stated that all workers, employees or others who used a workstation while at work were entitled to have a workstation which met the 'minimum specified requirement'.

Do all workstations have to be modified to meet these requirements?

If workstations do not already comply they will need to be modified to meet the conditions laid out in paragraph 1 of Schedule 1 to the DSE Regulations as follows:

- an employer shall ensure that a workstation meets the requirements laid down in the schedule to the extent that:
  - (a) those requirements relate to a component which is present in the workstation concerned
  - (b) those components have effect with a view to securing the health, safety and welfare of persons at work
  - (c) the inherent characteristics of a given task make compliance with those requirements appropriate as respects the workstation concerned.

In effect, employers have to ensure that all workers using DSE have a suitable environment in which to work, have the necessary equipment to work safely and that the tasks they do are managed effectively so as not to create health and safety issues.

What are the main areas to pay attention to when carrying out a workstation assessment or risk assessment?

Each workstation should be assessed with the following in mind:

- adequate lighting
- adequate contrast — no glare or distracting reflections
- distracting noise minimised
- leg room and clearances to allow postural changes
- window covering if needed to minimise glare
- software — appropriate to the task, adapted to the user, no undisclosed monitoring of the user
- screen — stable image, adjustable, readable, glare- and reflection-free
- keyboard — usable, adjustable, detachable, legible
- work surface with space for flexible arrangement of equipment and documents, glare-free
- chair — stable and adjustable
- footrest, arm/wrist rest if users need one.

Are all display screen equipment users entitled to an eyesight test and a free pair of glasses?

No. Only those employees who are classed as 'users' under the Regulations are covered by the Regulation applying to eyesight tests.

An employee who is a user of DSE can request an eyesight test, as can anyone who is to *become* a user, and the employer has to arrange for one to be carried out.

If an existing user requests a test, an employer must arrange for it to be carried out as soon as practicable after the request and for a potential user, before they become a user.

The continual use of DSE or VDU screens may cause visual fatigue and headaches and corrective glasses may reduce the eye strain often experienced. There is no evidence yet available, however, that frequent use of display screen equipment causes permanent eye damage or creates poor eyesight. Users with pre-existing sight conditions may just become a little more aware of them.

Eyesight tests should be carried out by competent professionals and must consider the effects of working at DSE so the optician (or medical equivalent) will need to know that the eyesight test is for working at display screens, etc.

Once an existing user has had an eyesight test, they can request one at regular intervals. The employer should determine what this interval should be with the user of the equipment and should take advice from the optician or other expert.

Eyesight tests which detect short or long sight, eye defects, etc. are *not* the responsibility of the employer — they need only concern themselves with an eye test which addresses any safety or health issues with regaed to using display screens.

The employer must arrange for an eyesight test when requested to do so. This could be by having arrangements with local opticians or by having the eye testing carried out on the premises by mobile health surveillance units, etc.

The employer can make arrangements with only one local optician and employees will have no choice who they visit. Alternatively, employers can have employees use their own optician if they prefer. The important thing for the employer is that they must facilitate such eye sight tests if requested to do so.

Employers are not responsible for the costs of 'normal' corrective spectacles — these are at the employees'/users' own expense. But an employer is responsible for the cost of any 'special' corrective appliances which the optician has determined need to be worn by the user to prevent them suffering unnecessary eye strain while using a display screen. The user is only entitled to a basic pair of corrective spectacles necessary for them to continue to use the display screen safely. 'Designer' frames, special lenses, etc. are not the responsibility of the employer.

Employers may make a contribution towards the cost of other types of corrective spectacles if those spectacles include the 'special corrective' features needed for the DSE work.

What does the employer need to do in respect of the provision of training regarding the use of display screen equipment?

The Regulations are quite specific about the duties of employers to provide training and information to display screen users.

The employer has to ensure that 'users' and those about to become 'users' of display screen equipment receive adequate health and safety training in the use of any workstation on which they may be required to work.

Training should be provided before a new employee becomes a user of the equipment. The purpose of the training is to ensure that those who are (or will be) users know and understand the hazards and risk associated with using display screen equipment.

Training on DSE can be incorporated into general health and safety training or induction programmes as it is good practice for everyone to be aware of hazards and risks of all work activities.

It is important that any training programme addresses the steps needed to reduce or minimise the following risks:

- musculo-skeletal problems
- visual fatigue
- mental stress.

Managers of those using DSE also need to be trained in health and safety issues relating to DSE as they can have important influence over a key health hazard — mental stress. Managers must be aware of the legal need for users to take regular breaks away from the screen, for workstations to be ergonomically friendly, etc.

Users must be trained on how to use their display screen equipment effectively. They must know how to make their own personal adjustments to height, tilt of screen, contrast, etc. They must know when they can take breaks and what other tasks are expected.

As with all health and safety training, it is important for employers to have a record-keeping system so that they will be able to demonstrate, if called upon to do so, that their employees, users or operators received suitable and sufficient training.

Occupational health issues can take several years to manifest themselves and elements of musculo-skeletal injury may occur after an employee, user or operator has left the company.

Evidence of training can be useful to show that, as an employer, you fulfilled your statutory duties and that the employee or user was aware of the hazards and risks and knew what to do to control them.

Keep training records for at least six years — longer is preferable. Computerised records must comply with the Data Protection Act 1998.

What information do users have to be provided with?

Users of display screen equipment must be provided with adequate information about:

- all aspects of health and safety relating to their workstations
- the steps taken by their employer to ensure compliance with the Regulations.

Users and operators of DSE need to know about the risk assessments which have been undertaken, the hazards and risks identified and the control measures that the employer has put in place to reduce the hazards and risks.

In addition, users and operators need to know what procedures are in place for them to have eye-sight tests, the frequency of tests, the provisions for the purchase of 'special needs' spectacles, etc.

Information should be given on when breaks can be taken, what other tasks need to be completed during these times, when they should have training, etc.

Employers should not forget their general duties to *all* employees and others, in respect of information, instruction and training on all work activities as required under the Management of Health and Safety at Work Regulations 1999.

Do individual employees have to fill in a self-assessment form?

No, not legally but it is good practice and allows the employer to have an overview of all employees and the specific and individual concerns they have.

The health effects of using display screen equipment vary from user to user and not everyone reacts in the same way.

Display screen equipment assessments are best done individually and generic assessments are not ideal, although they may be able to set a standard and provide useful guidance to users.

When individuals have completed their self-assessment forms they should be collated, reviewed and actioned by the employer or a competent person.

Any individual who has indicated that they have a specific problem should be re-assessed by the competent person and individual control measures agreed.

What are some of the control measures which can be implemented by users themselves?

Practical tips for users include:

- getting comfortable
    - adjusting the chair

  - adjusting the screen angle
  - adjusting the seat height so that eyes are at the same height as the top of the screen and arms are horizontal to the keyboard
  - creating enough clear work space
  - removing obstructions and unnecessary equipment
  - adjusting the position of the keyboard, mouse, document holder, etc.
  - avoiding glare from lights or windows
  - creating space to move feet and legs; provide a footrest if necessary
  - sit comfortably in the chair — make sure it is adjustable
- keying in
  - provide a space in front of the keyboard
  - provide a wrist rest if necessary
  - keep wrists straight
  - do not 'bash' the keys
  - do not overstretch fingers
- using a mouse
  - sit up straight, do not slouch
  - move the keyboard out of the way
  - keep wrists straight
  - sit close to the desk
  - use a cordless mouse
  - support forearms on the desk
  - do not abuse the mouse — treat it lightly
- reading the screen
  - adjust brightness and contrast
  - clean the screen regularly
  - use text size on the screen as large as practicable
  - select colours which are easy on the eye
  - do not be afraid to change the screen format to your own preference
  - make sure that characters do not flicker and that text, etc. is sharply focused

- posture and breaks
    - move about and change position
    - move legs and feet
    - take a break and do something different, e.g. answer the phone
    - do not sit in the same position all of the time
    - take frequent short breaks rather than longer infrequent ones.

DISPLAY SCREEN EQUIPMENT – SELF-ASSESSMENT

Location/Department:

Name:

Date:

1. JOB DESIGN

How long is spent on computer per day?. . . . .

Does this include other roles, e.g. answering telephone, etc.? Yes/No

Comment: . . . .

Can breaks be taken freely? Yes/No

Comment: . . . .

2. WORKSPACE AND FURNITURE

Is there sufficient space (3.7 m^2)? Yes/No

Comment: . . . .

Is desk sufficiently large to allow comfortable arrangement of work? Yes/No

Comment: . . . .

Is desk height suitable? Yes/No

Comment: . . . .

Is there adequate light? Yes/No

Comment: . . . .

Is there excessive noise? Yes/No

Comment: . . . .

Is temperature comfortable? Yes/No

Comment: . . . .

Are there adjustable blinds to windows to prevent reflections? Yes/No

Comment: . . . .

3. EQUIPMENT

Does screen have adjustable controls? Yes/No

Comment: . . . .

Does screen tilt/swivel? Yes/No

Comment: . . . .

Is screen free from reflection/glare? Yes/No

Comment: . . . .

Are digits clear and defined, is screen free from flicker? Yes/No

Comment: . . . .

Is keyboard separate from screen? Yes/No

Comment: . . . .

Is it easy to use/non-reflective, etc.? Yes/No

Comment: . . . .

Is there adequate space in front of keyboard to support hands and arms? Yes/No

Comment: . . . .

Is a document holder required? Yes/No

Comment: . . . .

4. CHAIR

Is chair fully adjustable? Yes/No

Comment: ....

Is it stable? Yes/No

Comment: ....

5. OPERATOR

Does operator know how to adjust chair to suit them? Yes/No

Comment: ....

Does operator know how to adjust display and position of screen to suit their needs? Yes/No

Comment: ....

Are they aware of associated risks? Yes/No

Comment: ....

Are they encouraged to take regular breaks? Yes/No

Comment: ....

Has information been made available for them regarding visual display screen use? Yes/No

Comment: ....

Signed (auditor) ....

Date ....

Signed (subject)..

Top tips

- Assess all workstations in the organisation.
- Review all equipment for comfort, ease of use etc.:
  - chairs
  - screens
  - keyboard
  - mouse/trackball
  - lighting/glare
  - environment
  - work/schedule demands
  - software.
- Record information in Risk Assessments.
- Decide on control measures to reduce hazards and risks.
- Offer eyesight tests to users — do not wait for them to ask.
- Determine what level of corrective spectacle you will pay for or towards.
- Introduce a comprehensive training programme and give out good levels of information.
- Keep training records for as long as possible.

Case study

Two women employed as data processing clerks sued British Telecom for repetitive strain injury.

The system of work they were forced to adopt involved long hours at the display screen and keyboard, keying in data at high speed. Incentives were given to work faster.

Furniture was not chosen with ergonomics in mind and chairs, etc. were not adjustable. No information was given to employees regarding best working posture and practices.

Both developed painful and ongoing RSI symptoms which prevented them both from working.

RSI was a developing industrial disease and information was available about it at the time of the incident (early 1990s).

The two operatives mounted a civil claim for damages and were successful.

The judge held BT liable for their injuries as he found that their injuries were purely a result of their work activity.

BT ought to have taken steps to correct the employees' postures and to have provided proper furniture and workstations, even though they may not have been fully aware of RSI.

9

Work equipment and work practices

What needs to be assessed in respect of work equipment?

The Provision and Use of Work Equipment Regulations 1998 requires that equipment provided by an employer for use at work is constructed or adapted so as to be suitable for the purpose for which it is provided and is used only for operations for which and under which conditions it is suitable.

In addition, the work equipment must be selected with regard to the working conditions, use, hazards and risks, etc. existing on the premises, and must be selected with regard to the health and safety of persons using it.

So, work equipment really needs to be subjected to a risk assessment, in particular:

- what the equipment is
- how it works
- where it will be used
- who will use it
- what the hazards and risks are
- what control measures are needed
- what the residual risks are
- what training operatives will need.

Employers must protect employees from any dangerous parts of machinery.

What are some of the hazards and risks associated with using work equipment?

Work equipment is a major cause of injury and has contributed to fatalities, major injuries, 'over three day' injuries, lost time and near miss incidents.

Failing to use equipment safely increases the risks of injury.

Hazards include:

- entanglement
- entrapment
- electricity
- hot surfaces
- impact damage
- excessive noise
- fire or explosion.

Risks include:

- death
- serious injury
- amputation of limbs
- electric shock
- musculo-skeletal problems
- industrial disease, e.g. deafness.

When should a risk assessment be carried out for work equipment?

A risk assessment is essential when new equipment is to be brought into the workplace, when equipment is to be changed or adapted and when the work process changes.

A risk assessment should also be carried out if the use of the equipment will affect people other than those at work.

Under the Dangerous Substances and Explosive Atmospheres Regulations 2002, a risk assessment must be carried out if any work activity or use of equipment involves or will involve an explosive atmosphere, e.g. the use of electrical equipment in a flammable environment.

What are some of the hazards, risks and control measures needed in respect of lifting equipment?

The Lifting Operations and Lifting Equipment Regulations 1998 require that lifting equipment is inspected by a competent person either:

- annually if only goods are lifted
- twice a year if people are lifted.

The Lifting Regulations also require lifting operations to be assessed to ensure that they are safe and suitable for the tasks in hand.

Lifting equipment includes lifts, lift cars and cages, lifting devices, hoists, block and tackles, ropes, chains, eye bolts, wheelchair lifts, scissor lifts, cradles, cherry pickers, etc.

Some common hazards to watch out for are:

- mobile cranes might overturn if not properly set up with stabilisers
- lifting machinery failing through mechanical fault, damage or overloading

- dangerous parts of machinery (e.g. closing traps), especially with scissor lifts
- trapping risks between the lifting machinery and overhead structures
- overhead electrical lines or cables close to lifting machinery.

Typical control measures are:

- maintaining lifting equipment in a safe condition
- ensuring that lifting equipment is examined, tested and certificated in accordance with statutory requirements
- keeping up-to-date records
- clearly marking the Safe Working Load (SWL) on all equipment
- always visually inspecting lifting equipment prior to use
- using the correct equipment for the task
- not exceeding the SWL
- ensuring that lifting equipment is only used, inspected and maintained by a competent person(s)
- using hydraulic hoists which are supplied by a pre-vetted contractor
- ensuring that contractors supply a risk assessment which applies to the task in hand
- considering medical review for fitness to work at heights.

Are hand tools considered equipment under the Provision and Use of Work Equipment Regulations 1998?

Yes. Hand tools can cause a number of injuries if not used correctly and, although they may be small, they can still be dangerous.

A risk assessment should consider the hazards, who could be harmed, how, what needs to be done to control the risks, etc.

Hand tools could include knives, hammers, chisels, screwdrivers, woodworking tools, maintenance kits, electrical appliance testers, etc.

Common hazards include:

- handles in poor condition which can cause hand injuries
- mushroomed heads on steel chisels which can cause fragments to be ejected
- blunt cutting edges which can require excessive forces to be used
- improper use of knives, e.g. cutting towards the body
- hand tools which are poorly modified or used for purposes for which they are not designed.

Typical control measures would be:

- providing good quality tools and maintaining them in good condition
- ensuring that users are adequately trained and competent
- providing special purpose tools where necessary, e.g. parcel knives for opening packages
- using sharp knives that are fitted with shielding or retracting blades where possible.

What are the hazards and risks to be assessed with regard to workshop equipment?

Workshop equipment will include lathes, milling machines, circular saws, power presses, grinding machines, drilling machines and a whole range of other industrial type equipment, e.g. abrasive wheels.

Woodworking machinery and power presses are known to be particularly high risk pieces of work equipment and need special attention in respect of health and safety while they are used by people at work.

Power presses must be inspected by competent persons and records must be kept on a *daily* basis to ensure that they are safe to use. In addition, power presses need to be thoroughly examined by a competent person at specified intervals.

Common hazards with workshop equipment include:

- inadequate guarding allowing contact with dangerous parts, e.g. rotating parts, cutting blades, traps between closing tools
- improper access into machinery cabinets by unauthorised staff, e.g. fault finding
- ejection of materials from machinery
- swarf and dust allowed to build up
- controls that are poorly identified and positioned
- harmful substances and dusts emitted during machining.

Typical control measures include:

- providing and maintaining suitable guarding
- ensuring that machines are used only by trained, competent people
- providing shields, guards and protective equipment where necessary
- leaving machines fit for next use
- clearly identifying and conveniently positioning machine controls
- providing and using good general ventilation and local exhaust ventilation.

What are the hazards and risks to be assessed with regard to pressure systems?

The Pressure Systems Safety Regulations 2000 apply to pressure vessels and systems and are designed to ensure that potentially dangerous and explosive equipment is used and maintained in a safe condition.

The hazard from a pressure vessel is high — it could explode and cause mass injuries, fire and great consequential damage. However, provided that a pressure system is well maintained the risk could be quite low.

Pressure systems include:

- boilers
- steam heating systems
- air compressors
- pressure cookers
- steam ovens
- heat exchangers
- process plant and pipes
- pressure gauges and hoses
- refrigeration plant.

Common hazards include:

- poor equipment and/or system design
- poor maintenance of equipment
- unsafe systems of work
- operator error, poor training or supervision
- incorrect or poor quality installation
- inadequate repairs or improper modifications.

Typical control measures are:

- providing safe and suitable equipment
- ensuring that the operating conditions are known, e.g. what fluid or gas is being contained, stored or processed
- checking that the process conditions are known, such as the pressures and temperatures
- ensuring that the safe operating limits of the system are known (and those of any equipment directly linked to or affected by it)
- ensuring that there is a set of operating instructions for all of the equipment and for the control of the whole system (including emergencies)

- giving appropriate employees access to these instructions and properly training them in the operation and use of the equipment or system
- fitting suitable protective devices and keeping them in good working order at all times
- establishing that checks are made to ensure that the protective devices function properly and that they are adjusted to the correct settings
- fitting warning devices, which are noticeable either by sight or sound
- ensuring that the system and equipment are properly maintained, taking account of its age, the environment and its use
- when protective devices have to be isolated for maintenance, making alternative arrangements to ensure that safety levels are not exceeded without detection
- establishing a safe system of work to ensure that maintenance work is carried out properly and under suitable supervision
- giving appropriate training to everybody operating, installing, maintaining, repairing, inspecting and testing pressure equipment
- checking that all persons carrying out the work are competent
- examining the equipment and ensuring that a written scheme of examination has been prepared by a competent person.

What are the requirements for working in confined spaces?

The Confined Spaces Regulations 1997 require employers to avoid the need for working in confined spaces unless absolutely necessary.

If work is required and there are no alternatives, employers must conduct a risk assessment and introduce suitable and sufficient control measures to reduce the risk to as low a level as possible.

A confined space has two defining features:

- it is a place which is substantially enclosed
- there will be a reasonably foreseeable risk of serious injury from hazardous substances or conditions within the space or nearby.

Examples include:

- ducts
- vessels
- culverts
- boreholes
- bored piles
- manholes
- sumps
- excavations
- inspection pits
- sumps
- building voids
- enclosed rooms
- cellars (some)
- inside machinery, autoclaves, etc.
- sewers
- silos.

Some confined spaces may be open-topped (e.g. tanks and vats), or inadequately ventilated structures (e.g. internal rooms).

The priority when carrying out the risk assessment is to identify the measures needed so that entry into the confined space can be avoided.

If this is not reasonably practicable, the employer must assess the risks connected with persons entering or working in the space and also the risks to others who could be affected by the work.

The person carrying out the risk assessment must be competent to do so, i.e. must understand the hazards and risks involved, be experienced and familiar with the relevant processes, plant and equipment and be competent to devise a safe system of work.

What are some of the hazards associated with confined space working?

Common hazards include:

- people entering confined spaces being overcome by fumes or gases already present
- people being overcome by fumes or gases which accumulate in the space after the person has entered
- people being affected by a lack of oxygen in the atmosphere of the space
- entry and exit from the confined space may be difficult.

One of the most common hazards associated with confined spaces is the 'multiple fatality' effect which occurs when other workers and rescuers enter the confined space and are overcome by the same fumes or gases which have affected those who entered first.

What are common control measures for working in a confined space?

The most effective and accepted control measure is a safe system of work operated under a 'Permit to Enter' system.

A Permit to Work or Enter will:

- outline the hazards
- describe the work process
- list the names of operatives
- define the hazards that are present
- list the checks that have been made
- list the PPE and safe systems of work to be followed
- limit the work or entry duration time
- stipulate competency for individuals
- stipulate training needs

Case study

The HSE reported in March 2003 that there had been four deaths involving confined spaces in four weeks.

Three deaths involved oxygen deficiency, and the other, a flammable liquid.

The deaths occurred in two separate incidents.

The fatalities occurred soon after entering the confined spaces and indicated that no matter how brief an entry into a confined space it can still be lethal.

Low oxygen levels have been attributed to the process of rust formation within a previously sealed vessel and the use of an inert gas in the welding process.

Hazards of a confined space include:

- oxygen deficient atmospheres
- toxic gases, fumes and vapours
- danger from flooding or other liquids entering the space
- the flow of solid materials such as grain in silos.

- name the competent person
- list the control measures to be taken
- list the checks to be made prior to handback.

A Permit to Enter will constitute a safe system of work if carried out properly.

Other control measures include:

- ventilating the confined space
- opening up covers, etc.
- testing the atmosphere for gases, fumes, etc.
- purging the confined space
- restricting the use of chemicals, etc.
- restricting the use of tools
- stipulating the minimum or maximum number of operatives
- stating protective clothing to be worn
- stating protective equipment to be used
- stipulating training, information and instruction to be given.

Effective emergency procedures and rescue systems *must* also be developed and implemented, including an operative who stays as ‘top man’ so as to be able to raise the alarm.

PERMIT TO ENTER A CONFINED SPACE

Project number . . . .

Date. . . . .

Location . . . .

Description of work . . . .

. . . .

. . . .

	Yes	**No**	**Does not apply**
Protective equipment			
Safety line and harness			
Goggles/face shield			
Head protection			
Protective clothing			
Hearing protection			
Fire extinguisher			
Respiratory protective equipment Type:____________			
Other			

	Yes	No	Does not apply
Questions which must be answered			
Have atmosphere tests been conducted?			
Document review and approval completed?			
Lockout/tagout procedure required?			
Job safety analysis required?			
Proper tools and equipment available?			
Emergency/explosion-proof lighting available?			
Oxygen test before entry?			
Explosimeter test before entry?			
Standby to be posted outside?			
Employees thoroughly briefed on the hazards and safe methods of doing the job?			
Hot Works permit?			
Adequate ventilation?			
Are signs properly posted?			
Is there a sign in/out log?			

Approvals superintendent . . . .

Safety representative . . . .

Standby . . . .

Note: A copy of this permit shall be posted at the worksite.

What are some of the common hazards, risks and control measures for working at heights?

New Regulations on working at heights are proposed by the HSE to come into force during 2004/2005 and these Regulations will be based on the need to carry out a risk assessment for working at heights, as defined under Regulation 3 of the Management of Health and Safety at Work Regulations 1999.

Working at heights includes:

- using ladders
- using step ladders
- using access cradles
- working from scaffolding
- working from platforms
- using tower scaffolds
- using mobile elevating work equipment
- working on roofs.

Hazards also include falling, falling objects, slipping or tripping, collapsing scaffolds, etc.

Common hazards include:

- ladders slipping on smooth flooring (e.g. lino)
- damaged rungs or rungs slippery with grease or mud
- metal ladders used for electrical work
- over-reaching from a ladder
- ladder not being footed by another person, or tied or lashed at the top
- an unsuitable ladder being used (e.g. too short)
- tools and/or materials being dropped by persons using them at a height
- failure of temporary fastenings, ropes or cables securing equipment
- materials being dislodged from high workplaces, e.g. by strong wind

- people injured by objects dropped by person working on ladder
- general access scaffolds not properly constructed, e.g. footings not stable, inadequate ties to buildings, insufficient bracing
- guard-rails and toe boards missing
- scaffold boards and access ladders not secured
- scaffold not inspected
- bracing and stabilisers missing from towers
- wheels not locked on moveable towers.

Typical control measures include the following:

- where possible, stable work platforms are used (e.g. tower scaffolds, etc.) instead of ladders
- workplaces are redesigned to obviate the need for ladders or steps
- ladders are used which suit the purpose (e.g. of the correct type or class)
- a ladder register is kept and ladders are regularly inspected and maintained
- stabilisers are used where practicable, or securing ties or a second person to foot ladders
- non-conducting ladders are used in electrical work
- safety bonds are provided to support overhead rigged equipment in case the main fitting fails
- work is prohibited overhead while persons are working below
- toe boards and scaffold nets are provided at tower and scaffold edges
- all materials kept on towers and scaffolds are tied down or netted
- lanyard is used to secure hand tools used over the edge of a working platform or cage
- barriers are provided to prevent persons passing underneath temporary overhead work
- all scaffolding is erected by trained and competent persons

- general access scaffolds are inspected by a competent person after erection and a certificate is obtained before handover
- general access scaffolds are inspected weekly when they are in use or after violent weather or following any changes to the scaffold structure
- wheels are locked or braked before a tower is used
- ladders or height extenders are not used on the platform of a tower
- towers are not moved while a person or heavy equipment is on them
- a safe means of access (e.g. ladder) is provided on the narrowest side of the tower and inside the tower structure
- the working platform is fitted with suitable guard-rails and toe boards
- work is not permitted on fragile materials.

10

Persons with special needs

As an employer, what are some of the common risks to new or expectant mothers which I need to be aware of?

There are certain aspects of pregnancy which can be exacerbated by various work activities and it is sensible to be aware of these so that special attention can be paid during the risk assessment process.

- Morning sickness and headaches:
  - consider early shift patterns
  - avoid exposure to nauseating smells
  - avoid exposure to excessive noise.
- Backache:
  - consider excessive standing
  - consider posture if sitting for prolonged periods
  - consider manual handling tasks.
- Varicose veins:
  - consider standing for prolonged periods
  - consider sitting positions and options for footrests.
- Haemorrhoids:
  - consider working in hot conditions.

- Frequent visits to the toilet:
  - consider location of work area in relation to WCs
  - consider ease of leaving the workstation
  - consider whether software records absences
  - consider confined or restricted working space
  - consider ease of leaving the job quickly.
- Increasing size:
  - consider use of protective clothing or uniforms
  - consider work in confined spaces
  - consider manual handling tasks.
- Tiredness:
  - consider whether overtime is necessary
  - consider shift patterns
  - consider whether evening work or early morning work is necessary
  - consider flexible working hours.
- Balance:
  - consider housekeeping to avoid obstacles
  - consider unobstructed work areas
  - consider specifically slippery floor surfaces
  - consider exposure to wet surfaces.
- Comfort:
  - consider too tight clothing
  - consider temperatures of the work area — too cold, too warm
  - consider 'over-crowding' with fellow employees
  - consider whether tasks need to be done at too great a speed.
- Stress:
  - consider anything which could cause an expectant employee to become anxious about any working conditions.

What steps are involved in completing a risk assessment?

As with all risk assessment procedures, a planned approach is best.

Risk assessment is best broken down into steps or stages.

It is best to begin risk assessment for new or expectant mothers before any employee becomes pregnant — an employer is well advised to assess job activities which *could* cause problems for new or expectant mothers.

Stage 1: Initial risk assessment

Take into account any hazards or risks from your work activities (or those of others) which could affect employees of childbearing age.

Risks include those to the unborn child or to a child of a woman who is breastfeeding.

Look for hazards which could affect all female employees, not just those who are pregnant.

Physical hazards

- Movement and posture
- Manual handling
- Noise
- Shocks or vibrations
- Radiation
- Impact injuries
- Using compressed air tools
- Working underground.

Biological hazards

- Working with micro-organisms which can cause infectious diseases.

Chemicals, gases and vapours

- Toxic chemicals
- Mercury

- Pesticides
- Lead
- Carbon monoxide
- Medicines and drugs
- Veterinary drugs.

Working conditions

- Stress
- Working environment
- Temperatures
- Ventilation
- Travelling
- Violence
- Passive smoking
- Working with DSE
- Working hours
- Use of personal protective equipment
- Use of work clothes or uniforms
- Lone working
- Working at heights
- Mental and physical fatigue
- Rest rooms, work breaks, etc.

Consider whether any of the above can harm any employee, but particularly new or expectant mothers.

The requirement to control many of the above hazards is contained in specific health and safety regulations, e.g. Manual Handling Operations Regulations 1992.

Decide who is likely to be harmed, how, when and how often, etc. Remember that new or expectant mothers may be less tolerant to hazards than other workers and so the degree of control needed to eliminate or reduce the risks may be greater than would normally be expected.

Consult your employees and inform them of any risks identified by the risk assessment. In particular, advise all female employees of childbearing age of any risks which may affect them. Advise them of the steps that are to be taken to reduce the risks.

EXAMPLE RISK ASSESSMENT – GENERIC RISKS FOR A PREGNANT EMPLOYEE

Task/Activity

Working while pregnant, breastfeeding or as a new mother — general office work, lifting loads, sitting at desks and driving

Who is at risk from the activity?

Pregnant and breastfeeding women and those that have given birth in the past six months (new mothers)

What are the hazards (dangers)?

Excessive strain	Overstretching/twisting
Heat	Tiredness
Chemicals	Slips/trips and falls
Lifting	VDU screens/workstations

What are the potential outcomes from the hazards?

Fainting	Misjudgement
Chemical exposure	Slips/falls
Back strains	Personal injury
Damage to unborn baby	Repetitive movement
Eyestrain/headaches	

What is the likelihood of the risk occurring?

HIGH

MEDIUM

LOW

How do we currently control these risks?

- Allow regular breaks to relieve muscle strain and tiredness which may add to misjudgement.
- Ensure environment is well ventilated and drinking water provided and regular breaks taken.
- Do not lift excessive weights. Always get help with lifting. Ensure training is given.
- Read COSHH data prior to chemical use.
- Wear appropriate protective equipment for chemicals used.
- Minimise activity in order to avoid stress, excessive stretching, twisting and strain to the body.
- Keep floor surfaces slip/trip hazard free and in good condition.
- Ensure line manager undertakes review of pregnancy and activities by using form in health and safety policy to address site-specific and person-specific duties and actions.
- Take regular breaks from driving — request back support for car if necessary.
- Avoid undertaking long journeys or driving where regular breaks are not possible.
- Take regular breaks from VDU screen — 10 minutes in every hour undertaking different activities.
- Provide footrest at VDU if necessary.
- Ensure VDU self-assessment undertaken (refer to health and safety policy).
- Availability of non-smoking rest area, with seating and no smoking at the bar rule to control smoke level at the bar.

References

Health and Safety Policy

Assessment form to support pregnancy risk assessment

What else can we do/what else is required?

- Monitor activities of employee.
- Do not work if feeling unwell — advise manager of this.
- Do not continue to work if unable (i.e. near birth).

Who prepared the risk assessment and when?

Who needs to know about these findings?

Managing Director

Managers

All staff (all females of childbearing age)

What is a site-specific risk assessment and why is this necessary?

Generic risk assessments are invaluable tools in helping to assess general hazards and risks in respect of pregnant employees or nursing mothers.

However, everyone is different and pregnancy affects women in different ways. So, they cannot all be treated the same.

'Person-specific' risk assessments will ensure that employers can demonstrate that they took due regard of the hazards and risks to individuals.

A specific risk assessment should be completed each trimester as the effects of hazards and risks will vary — something that was acceptable in the first three months may not be quite so acceptable when the employee is six months pregnant.

RISK ASSESSMENT: ADDITIONAL SITE-SPECIFIC INFORMATION

Employee name:

Expectant/New Mother
(Please highlight applicable category)

Premises name:

Please indicate stage of pregnancy:

1–3 months ___ 4–6 months ___ 7–9 months ___

New mother ___ Breastfeeding mother ___

NB This form is to be completed for all stages as indicated above.

In addition to the general pregnancy risk assessment available, this specific pregnancy assessment reviews the working conditions, environment and medical status of the pregnant employee through each stage of the pregnancy and return to work after giving birth.

The Line Manager must complete this form *with* the pregnant employee at each of the above stages. The guidance contained within the assessment form will give recommendations and indications of action to be taken where hazards may be identified.

Manual handling				
If the answer is 'yes' to any of the tasks/activities detailed below, give exact details of tasks under 'detail'. Also, if the answer is 'yes', where 'action to be taken' details 'minimise task', consider *with* the employee, what is reasonable, what they are capable of and comfortable with. They may usually be able to lift and carry a *limited* amount of weight. Also, movements such as 'reaching upwards' or 'twisting' will need to be reviewed but so long as they are not excessive, and do not involve, for example, significant risk of injury or lifting weight while doing so, this sort of activity will normally be reasonable to expect.				
	YES	**NO**	**Detail (if yes)**	**Action to be taken (if yes)**
Is there risk of manual handling injury from:				
Lifting filled crates/other heavy goods				Eliminate task
Carrying food deliveries				Minimise task
Carrying stock				Minimise task
Moving furniture				Minimise task
Pushing/pulling items				Minimise task
Reaching upwards				Minimise task
Twisting movements				Minimise activity where practical

Working environment

Consider whether any such exposure to the factors below would mean an increased risk of injury. For example, working behind the bar, the risk of slips *may* be elevated but consider measures to help prevent and to deal with spillages. Are they adequate, do they need improvement? Detail any actions to be taken as a result of your assessment.

	YES	**NO**	**Detail (if yes)**	**Action to be taken (if yes)**
Is there any risk of injury from:				
Slips, trips and falls — *consider trip hazards, risks of falling.*				
Cramped working space — *any confined spaces, cramped working conditions that the pregnant employee may have to enter, e.g. an area where they have to bend down/slouch to work.*				
Excessive cold — *consider the walk-in freezer for example — will the employee have to be in there for any significant period of time and, if so, what can be done. Can they limit time spent in areas such as freezer, can they wear protective, warmer clothing.*				
Excessive heat — *consider areas such as the kitchen — can regular breaks be taken, is drinking water available, ventilation working?*				

Working environment *continued*				
	YES	**NO**	**Detail (if yes)**	**Action to be taken (if yes)**
Excessive travelling distances — *does the employee have to walk very far from work areas? Does she have to travel far to work by public transport? Contact Personnel department — can she be transferred to another premises for a temporary period or can she stay at a local hotel if she finishes a shift late at night, e.g. past 11 pm?*				
Continual use of stairs — *does the employee have to go up and down stairs at work? Is this excessive and likely to lead to increased fatigue? Can it be avoided? What can be done?*				
Exposure to excessive and consistent tobacco smoke — *is ventilation working, is there always a non-smoking area available? If problems are identified, detail and decide on action.*				
Exposure to excessive noise — *if premises has very loud music at times when pregnant employee is working, is this excessive? Is ventilation working in the premises other mechanical equipment without too much noise? Detail any action needed.*				

Working time/Activity				
	YES	**NO**	**Detail (if yes)**	**Action to be taken (if yes)**
Is there a risk of injury, ill health or fatigue from:				
Standing for long periods, i.e. over 2 hours without breaks				Provide a seat/stool to sit down at quieter periods.
Working excessive shifts (i.e. not allowing an 11-hour break in 24-hour shift work)				Contact Human Resources department to reduce shifts.
Working more than 48 hours				Reduce hours to below 48 hours per week. Contact Human Resources department.
Lack of seating				Provide seated, non-smoking rest area for breaks.
Lack of breaks (should be at least 40 minutes in every six hours)				Provide additional breaks — contact Human Resources department.

Workstations/VDU stations				
Review the use of any VDU by the pregnant employee — refer to the relevant section in the Health and Safety policy. Detail any necessary action to increase comfort if needed and allow regular breaks away from the screen (at least 10 minutes in every hour).				
	YES	**NO**	**Detail (if yes)**	**Action to be taken (if yes)**
Is there risk of injury, ill health or fatigue from:				
Long periods at the workstation (i.e. over 2 hours at any one time)				Increase breaks away from screen to 10 minutes in every hour doing different tasks.
Glare from screens				Fit screen glare guard.
Poor work position				Improve with seating, etc.
Inadequate space (i.e. should be enough space to sit and work comfortably)				Contact Human Resources department.
Lack of breaks				Contact Human Resources department.

Use of chemicals, etc.				
Check all chemicals used are those approved in the COSHH manual. Review the employee's contact with body fluids spillages — can they be avoided to an extent, i.e. can they avoid clearing body fluid spillages, etc.				
	YES	**NO**	**Detail (if yes)**	**Action to be taken (if yes)**
Is there any exposure to:				
Chemicals/substances not in COSHH manual				Contact Human Resources department.
Fumes, excessive smells, dust, etc.				Contact Human Resources department.
Biological agents (e.g. body fluids)				Avoid clearing vomit, urine, blood, etc. with spill packs.

Night working				
If the employee is working during the evening/night shifts, e.g. past 8 pm, consider what can be done to give extra breaks/rest. Can she sit down during quieter periods, is there access to a seated, non-smoking rest area? How will she get home after a night shift? Consider options as detailed in 'Working environment'.				
	YES	**NO**	**Detail (if yes)**	**Action to be taken (if yes)**
Will the employee be working at night, i.e. on a night shift (past 8 pm)?				
Will the employee be working until 11.00 pm, 12.00 pm or 1.00 am?				

Wherever 'yes' has been indicated above, make sure sufficient detail has been recorded. If you need to provide further detail, include it in this section.

Other issues

Describe any other issue, which may affect the individual's health and safety due to being either an expectant or new mother, or because of an existing medical condition. For example, pre-existing medical conditions that may affect a woman during pregnancy or afterwards include:

Diabetes

Any heart condition

Joint/bone conditions, e.g. osteoporosis, previous injuries causing pain or discomfort

Previous miscarriage

THE PREGNANT EMPLOYEE IS RESPONSIBLE FOR CONTACTING THE HUMAN RESOURCES DEPARTMENT IF SHE FEELS THAT THERE MAY BE A MEDICAL CONDITION THAT SHOULD BE RECORDED, WHICH SHE HAS NOT RAISED DURING THE ASSESSMENT.

Signed (employee): ....

Date: ....

Signed (Line Manager): ....

Date: ....

Retain copy for personnel file and copy to Human Resources department.

What needs to be considered in the risk assessment for young workers?

Firstly, the risk assessment must be completed *before* the young person starts work or work experience.

Each young person must be told what the hazards and risks are and must have control measures explained, etc.

The risk assessment must:

- consider the fact that young people are inexperienced in work environments
- consider that young people are physically and mentally immature
- consider their lack of knowledge in work procedures
- consider that they are inexperienced in perceiving danger
- consider that their literacy skills may be less than ideal
- consider all the control measures necessary to reduce or eliminate the hazard
- consider that personal protective equipment, etc., if identified as the control measure, may be sized for adults and may not therefore be 'suitable and sufficient'
- be kept up to date
- be relayed to and discussed with the young person
- identify training needs for the young person
- consider the tools and equipment they will use — whether there is an age restriction, e.g. on dangerous machines, cleaning, etc.
- consider the layout of the workplace
- consider the environmental hazards
- consider any hazardous substances in use.

Are there any risks which young people cannot legally be exposed to?

Young people under the age of 18 years must not be allowed to do work which:

- cannot be adapted to meet any physical or mental limitations they may have
- exposes them to substances which are toxic or cause cancer
- exposes them to radiation
- involves extreme heat, noise or vibration.

If young people are over the age of 16 years, however, they may be able to undertake or be exposed to the above tasks and hazards if it is necessary for their training and if they are under constant supervision by a competent person.

Children below the school-leaving age must *never* be allowed to undertake the tasks or be exposed to the hazards listed above.

What training and supervision do young people require under health and safety law?

Young people need training when they start work — *before* they undertake any work activity, process or task. They must be trained to do the work without putting themselves or others at risk.

It is important to check that young people have understood the training and information they have been given.

- Do they understand the hazards and risks of the workplace?
- Do they understand the basic emergency procedures, e.g. fire evacuation, first aid, accident reporting?
- Do they understand the control measures in place to eliminate or reduce risks?
- Do they understand their responsibilities as employees not to interfere with safety equipment, not to fool about, etc.?

Young people must be regularly supervised by competent people, i.e. those who understand that a young person may inadvertently put themselves or others at risk because they do not know any better or cannot perceive the risk.

11

Personal protective equipment

What is personal protective equipment?

The Personal Protective Equipment at Work Regulations 1992 define personal protective equipment (PPE) as:

> all equipment (including clothing affording protection against the weather) which is intended to be worn or held by a person at work and which protects him from one or more risks to his health or safety.

Personal protective equipment therefore includes, but is not limited to:

- safety helmets
- ear protection
- eye protection
- gloves
- safety shoes or boots
- high visibility jackets
- respiratory masks
- breathing apparatus.

Ordinary working clothes (e.g. uniforms) are exempt from the Regulations unless they are worn for health or safety reasons, e.g. steel chain sleeved jackets.

What does the law require an employer to do in respect of personal protective equipment?

An employer has to carry out a risk assessment of the job tasks and activities which his employees have to undertake and must determine the control measures necessary to reduce hazards and risks to the lowest possible level.

If elimination of the hazard at source, engineering controls or a safe system of work will not reduce the hazard to an acceptable level, then the employer must provide the employee with appropriate PPE.

Personal protective equipment provided to employees must be free of charge.

Regulation 6 of the Personal Protective Equipment Regulations 1992 requires an employer to ensure that an assessment is made in respect of PPE needs. This must include:

- identifying risks not avoided by other means
- defining the characteristics required of the PPE
- comparing these with the characteristics of PPE available.

The assessment should be recorded and kept readily available in all but the simplest of cases.

Will the risk assessment carried out under the Management of Health and Safety at Work Regulations 1999 be sufficient to comply with the PPE Regulations?

Yes, provided the risk assessment has given due regard to the needs of PPE and has determined whether PPE is suitable.

The risk assessment under the Management of Health and Safety at Work Regulations 1999 will determine what health and safety measures are needed to comply with legal requirements.

If the conclusion is that PPE is required, another risk assessment is required under the PPE Regulations 1992 in order to determine if the PPE is suitable.

In reality, the two risk assessments are combined.

Remember that a risk assessment under the Management Regulations must be 'suitable and sufficient'. There is no such qualification for assessments under the PPE Regulations.

When does a risk assessment under the PPE Regulations 1992 have to be carried out?

An assessment has to be made before choosing any personal protective equipment which has to be provided.

What are the key requirements for PPE?

Many factors need to be considered when choosing personal protective equipment because choosing the wrong sort, or an ill-fitting type or an inappropriate specification may cause the employee to be exposed to greater hazards than having no PPE at all.

Often, people believe themselves to be adequately protected because they trust their PPE. Sometimes such trust is misplaced because the equipment is not providing protection.

In general, PPE must:

- be capable of adjustment to fit correctly
- be appropriate for the risks and conditions
- take account of ergonomic requirements and the state of the wearer's health
- prevent or adequately control the risks to which the wearer is exposed, without adding to those risks

- comply with appropriate product safety and other standards, e.g. CE marking
- be compatible with any other type of PPE being worn.

Who is responsible for monitoring and cleaning PPE?

The employer has a duty to maintain all PPE provided for his employees. PPE must be maintained in an efficient state, in efficient working order and in good repair.

Specific arrangements may be needed for:

- inspecting PPE
- maintaining PPE
- cleaning PPE
- disinfecting PPE
- replacing PPE
- examining PPE
- testing PPE.

The Company Safety Policy should state who is responsible for carrying out pro-active PPE inspection and testing procedures.

Defective and inappropriate PPE is often a greater hazard than no PPE.

It is appropriate at times to give 'users' some responsibility to check their own PPE, e.g. to ensure that face masks fit correctly. Employees who use PPE must know what to do to report that their PPE may be defective.

What records need to be kept?

A record of any risk assessment should be in writing where the employer has 5 or more employees but it is good practice to keep a record no matter how many employees there are in the company.

It is essential to keep a record of the assessment, i.e. why you chose the PPE you did, what its purpose was for, when it was issued, to whom, what training they have had and when the PPE should be inspected and replaced.

Most employers keep an employee record sheet so that they know who has what equipment and what to expect to be returned if an employee leaves the business.

What training does an employee have to have on the use of PPE?

Regulation 9 of the Personal Protective Equipment Regulations 1992 requires employers to give employees adequate and appropriate information, instruction and training on any PPE that they are expected to wear in order to safeguard them from hazards at work.

The training, information or instruction must include:

- why and when PPE must be used
- how to use it
- its limitations
- arrangements for its maintenance and/or replacement.

'How to use it' training may involve a practical demonstration on how to wear masks, breathing apparatus, goggles, etc.

Records of training must be kept. It is not sufficient merely to give employees PPE to wear or use. They *must* be given information on why to use it, how to wear it, etc.

Case study

Mrs W worked for the County Council as a cleaner. She was given rubber gloves to wear but did not really know when to wear them, why she needed to, etc. So she and her colleagues rarely wore them.

Mrs W used a range of detergents and cleaning chemicals over the years of her employment but, after some years, she started to develop eczema on her hands and wrists. Dermatitis was diagnosed and Mrs W's GP advised her to wear cotton gloves under her rubber gloves. She did this but the eczema got worse and spread to her face and other parts of her body. She gave up work on the advice on the medical profession.

Mrs W sued her employers for damages on the grounds that they had not protected her health and safety.

The Court found in favour of Mrs W and awarded her compensation.

The Court decided that it was not enough for the County Council to gives its employees personal protective equipment, i.e. gloves. In order to discharge its duty as an employer to provide a safe system of work, the Council should have warned the cleaners about the dangers of handling the cleaning chemicals without protection and should have instructed them to wear gloves at all times.

The Judge also concluded that the risks of dermatitis were well known from cleaning chemicals and the employers ought to have known of the risks. But the risks were not so well know that they would have been obvious to the staff without them receiving any necessary warning or instruction.

What are the responsibilities on employees to wear PPE given to them by their employer?

Employees have duties under the PPE Regulations 1992 to:

- use PPE in accordance with their training and instruction
- return PPE after use to the accommodation provided
- immediately report loss or damage to the PPE they use.

If an employee has had appropriate training, information and instruction and yet they persist in abusing or mis-wearing their PPE and they have an accident because the PPE did not protect them, they may be considered as contributing to their injuries by neglect and therefore not eligible for compensation, and/or they may be guilty of a criminal offence under Section 7 of the Health and Safety at Work Etc. Act 1974 (which requires employees to co-operate with their employers and not to jeopardise their own or others' safety).

As an employer, can I force my employees to wear the PPE I have provided?

Employers have a duty under Regulation 10 of the PPE Regulations 1992 to take all reasonable steps to ensure that PPE is properly used by employees.

Employers are therefore expected to take a proactive approach to enforcing their PPE rules and to introduce appropriate measures to ensure that the message is understood.

Many employers include the flaunting of PPE requirements as gross misconduct under employment contracts and deal with breaches as disciplinary issues.

What factors need to be considered when choosing PPE?

There are three distinct areas to consider when choosing PPE:

(1) the workplace
(2) the work environment
(3) the PPE wearer.

The workplace

- What are the hazards to be controlled?
- What machinery is in use?
- How many people work in the area?
- What are the known risks?
- What is PPE expected to do?
- Will people need to have movement when using their PPE; will they need dexterity to use equipment, etc.?

The work environment

- Will excessive temperatures be generated?
- Will ventilation be available?
- Is it to be used in a confined space?
- Could the PPE cause other hazards, e.g. prevent alarm bells from being heard?

The PPE wearer

Fit — what size is needed, does it feel comfortable to wear, is it adaptable to the wearer with adjustable straps, etc., does any physical feature affect its fit, e.g. beards and face masks?

Training — do users know what they have to wear and why, what the hazards and risks are, how to wear the PPE, how to inspect it before use, how to report defects, etc.?

Acceptability and comfort — will users wear it for prolonged periods, will it be too heavy or cumbersome? Will it slow them down so that 'piece work' may be affected? Does the user have choice?

Interference — will the PPE work with other PPE (e.g. goggles and masks), will it prevent other controls being used, e.g. will ear defenders prevent warning buzzers, etc. from being heard, will goggles restrict vision, etc.?

Management commitment — do employees see a culture of 'do as I do' and not 'do as I say but don't do'? Are management, including supervisors, leading by example?

What sort of PPE is necessary for different job tasks/activities?

Hearing protection

There are two main types of hearing protection — things placed in the ear canal and objects placed around the outer ear.

Ear plugs

- Fit inside the ear canal and impede the passage of sound energy to the ear drum.
- Made from glass down, rubber or foam.
- Are usually disposable.
- Allow some sound to the ear drum.
- Create hygiene problems if re-used.
- Are unlikely to produce a good fit.
- Come in varying sizes but could be too small or big for each individual ear.

Ear muffs

- Rigid cups that fit over the outer ear and are held in place with a head band.
- Are unlikely to be a tight fit.
- Allow some sound to the ear drum.
- Can be uncomfortable.
- Difficult to wear with glasses or goggles, etc.

Respiratory protective equipment

There are two broad categories — respirators which purity the air to be breathed in and masks which filter out particles.

Breathing apparatus purifies the air and draws uncontaminated air from an independent source to the wearer, e.g. from oxygen cylinders.

Face masks can be full face or half masks.

Dust masks prevent particulate matter from being breathed in by containing it in the filtering medium.

Eye protection

Eyes can be protected by goggles, spectacles or face masks.

Spectacles/glasses

- Used for protection against low energy projectiles.
- Do not protect against dust.
- Are easily displaced.
- Are usually incompatible with other PPE.

Goggles

- Used for protection against high energy projectiles.
- Protect against dust, etc. as they protect the whole of the eye and surrounding face.
- May mist up and prevent vision.
- May be uncomfortable and cumbersome.
- Protect against splashes.

Full face masks

- Provide full face protection.
- May be uncomfortable to wear.

Head protection

Head protection is usually one of two types.

Safety helmet

- Safety helmets must be worn on construction sites.
- Protect against impact damage.
- Deteriorate in sunlight.

Bump cap

- Provides less protection from impact damage.
- Easier to wear, especially in confined spaces.

Hand/arm protection

Gloves/gauntlets

- May be leather which affords good protection.
- May be cumbersome.
- May prevent good grip.
- May cause dermatitis.
- Are resistant to fluids.
- May be chain mail to protect against cuts.
- May be of rubber, PVC, nylon, cotton, latex or cloth.
- May corrode in certain substances.
- May be easily damaged.
- Can be uncomfortable to wear.

Safety footwear

Usually used where there objects could fall onto feet or where objects may pierce the soles of shoes, to offer protection in wet conditions, etc.

Safety boots/shoes

- May have steel toe caps.
- May have steel soles.
- Can be lightweight.
- Are often cumbersome.
- Protect against oils, etc.

Wellingtons

- May have steel toe caps.
- Protect against water and fluids.
- May not provide good grip.

Skin barrier creams

Used to assist in protecting hands from direct contact with fluids, cleaning chemicals, etc.

Other types of PPE

- Safety harness.
- Lanyards.

The eight requirements of PPE Regulations

1. PPE is to be provided as the last resort.
2. A suitability test is to be applied.
3. Compatibility between different PPE is to be ensured.
4. Maintenance and replacement schemes must be in place.
5. Places to store PPE to be provided.
6. Proper use to be made of PPE.
7. Users to report defects or loss.
8. Information, instruction and training to be provided.

PERSONAL PROTECTIVE EQUIPMENT RISK ASSESSMENT FORM

Name of company:

Assessment carried out by:

Job activity/type under review:

Name of operatives undertaking tasks:

Hazards associated with task/activity:

Personal protective equipment required:

Type	**Purpose**	**When to be worn/used**
Maintenance, inspection, cleaning procedures		
Training, information and instruction		
Review of assessment (date)		

PERSONAL PROTECTIVE EQUIPMENT RECORD SHEET

Name of employee:

Job occupation/title:

Main hazards associated with job tasks/activities:

Personal protective equipment issued:

Type	Condition	Date issued

Training information and instruction undertaken:

Subject	Date	Signed

Signed: ………………………… (employee)
Date: ………………………………………

Signed: ……………………….... (supervisor)

Date: ………………………………………

12

First aid

What is the main piece of legislation which covers first aid at work?

The Health and Safety (First Aid) Regulations 1981 set the standards for first aid at work.

The main scope of the Regulations covers:

Every employer must provide equipment and facilities which are adequate and appropriate in the circumstances for administering first aid to employees.

Employers must make an assessment to determine the needs of their workplace. First aid precautions will depend on the type of work and, therefore, the risk being carried out.

Employers should consider the need for first aid rooms, employees working away from the premises, employees of more than one employer working together and non-employees.

Once an assessment is made, the employer can work out the number of first aid kits necessary by referring to the Approved Code of Practice.

Employers must ensure that adequate numbers of 'suitable persons' are provided to administer first aid. A 'suitable person' is someone trained in first aid to an appropriate standard.

In appropriate circumstances the employer can appoint an 'appointed person' instead of a first aider. This person will take charge of any situation, e.g. call an ambulance, and should be able to administer emergency first aid.

Employers must inform all employees of their first aid arrangements and identify trained personnel.

What is first aid at work?

First aid at work covers the arrangements you must make to provide employees with adequate first aid attention while they are at work.

Employees may suffer injury or ill-health while at work, caused by a work activity or the work environment, or employees may become ill for other reasons while at work.

They must receive immediate emergency attention and, in serious cases, an ambulance must be called.

First aid at work is designed to save lives and to prevent minor injuries or incidents escalating into serious ones.

As an employer, what do I need to do?

The Health and Safety (First Aid) Regulations 1981 require employers to provide adequate and appropriate equipment, facilities and personnel to enable first aid to be given to employees if they are injured or become ill at work.

The Regulations set out some minimum first aid provisions to be provided on an employer's site as follows:

- a suitably stocked first aid kit
- an appointed person to take charge of any incident and the first aid arrangements
- first aid facilities to be available at all times when people are at work.

The key words regarding first aid are:

- 'adequate and appropriate'
- 'suitable and sufficient'.

In order to ascertain what appropriate first aid is required, employers will have to carry out a risk assessment of first aid needs.

What should an employer consider when assessing first aid needs?

The risk assessment process requires employers to consider the hazards and associated risks involved in work activities.

Small businesses will need only the simplest of risk assessments and basic first aid provision.

Larger businesses will need to consider the following.

Step 1

Consider what are the risks of injury and ill-health associated with your work practices and activities.

Step 2

Are there any specific risks which can be clearly identified, e.g.:

- working with hazardous chemicals
- working with dangerous tools
- working with dangerous machinery
- working with dangerous loads
- working with animals?

Step 3

Are there areas within the business where risks may be greater because of the environment and might these need additional first aid facilities, e.g.:

- research laboratories
- pathology laboratories
- hot working environments
- cold working environments?

Step 4

Consider the businesses record of accidents and injuries. What types have occurred? How serious have they been? Is there any evidence to show that extra precautions, etc. are necessary? Where and when did they happen, and why?

Step 5

How many people are employed on the site? How many are permanent, temporary, etc.? How familiar are they with procedures, processes, etc.?

Are there any young or inexperienced workers more likely to have an accident or suffer ill-health than the regular workforce? Does anything extra need to be done for people with disabilities?

Step 6

What are the buildings used for and how are they used, are they spread out, do employees work out of doors, can they access all parts of the building? Where might first aid provisions and facilities be located so that they are available to all?

Step 7

Do employees work outside of normal working hours, work overtime, work alone, etc.? How might they raise the alarm? Do employees travel frequently?

Step 8

Would emergency services be easy to summon? Could they gain access to the site or building if it were out of normal hours? Are work places accessible?

What are some of the steps an employer needs to take in order to ensure suitable first aid provision?

The assessment of the likelihood and frequency of injury and ill-health will determine what will need to be done to ensure suitable first aid provision, i.e. that emergency aid is provided.

First aid kits will need to be provided — should there be one in a central location or will it be preferable to have several smaller ones easily accessible to employees?

An appointed person or first aider will need to be appointed depending on the number of employees and severity of risk of injury, etc.

First aid room facilities may be needed depending on the number of employees.

Emergency procedures to call the emergency services will be needed.

Special consideration will be needed for people with disabilities.

The type of first aid equipment will need to be considered, e.g. will eye-wash stations be needed?

The workers and employees of other employers will need to be considered — who is providing first aid facilities for them, e.g. on construction sites?

What is an 'appointed person'?

An appointed person is someone who is not necessarily trained in emergency first aid treatment but who is appointed to take charge of an incident and call the emergency services, etc.

The appointed person usually also looks after the first aid equipment and ensures that it is adequately stocked, in the right location, etc.

Employers who employ up to 50 employees and whose business falls into the category of 'low risk', i.e. office environment, libraries, retail shops, etc. need only appoint an 'appointed person'.

There should also be 'reserve' appointed persons to cover for holidays and sickness.

What is a first aider?

A first aider is someone who has undergone training in first aid and holds a current first aid at work certificate. All first aider training courses must be approved by the HSE, so ask all providers for details of their HSE registration documentation, etc.

Employers who have over 50 employees usually need to appoint a trained first aider.

First aid training courses last four days and the certificate is valid for three years. After that, retraining is needed.

A trained first aider can administer first aid and the primary purpose is to prevent injuries from getting worse rather than to try to treat people or provide medical expertise.

How many first aiders or appointed persons does an employer need?

The Regulations do not set down hard and fast rules in respect of numbers of people to be appointed. It really is dependent on the type of work activity and the likelihood of injury.

The law requires the employer to make the assessment and, as long as that can be explained and justified as being suitable and sufficient, the law will be satisfied.

However, the Approved Code of Practice on First Aid at Work gives guidance on suitable numbers of appointed persons and first aiders.

Enough people should be nominated and trained so that absences can be covered.

A suggested ratio of appointed persons or first aiders is listed as follows.

Shops, offices, libraries	Less than 50 employees	1 appointed person
Shops, offices, libraries	50–100 employees	1 first aider
Shops, offices, libraries	Over 100 employees	1 first aider, plus an extra person per 100 employees
Food processing, warehouses	Fewer than 20 employees	1 appointed person
Food processing, warehouses	20–100 employees	1 first aider for every 50 employees
Food processing, warehouses	More than 100 employees	2 first aiders plus an extra person for every 100 employees
Construction sites, industrial sites, manufacturing, spray shops, chemical industries, etc.	Less than 5 employees	1 appointed person
Construction sites, industrial sites, manufacturing, spray shops, chemical industries, etc.	5–50 employees	At least 1 first aider
Construction sites, industrial sites, manufacturing, spray shops, chemical industries, etc.	More than 50 employees	A first aider for every 50 employees

What should be in a first aid kit?

There is no legal list of items which should be in a first aid box, although there is guidance in the First Aid Approved Code of Practice.

The contents of a first aid kit really depend on the risk assessment for first aid facilities which every employer must complete.

First aid kits should *not* however, contain any medicines, e.g. aspirin or paracetamol, because no one is trained to issue medicines and the recipient could be allergic to the substance, etc.

The contents of a first aid kit are relatively straightforward and the following would be sensible contents for a low risk work environment:

- 20 individual sterile adhesive dressings (plasters) of varying sizes
- 2 sterile eye pads
- 4 sterile triangular bandages
- safety pins
- 6 medium-sized sterile wound dressings
- 2 large-sized sterile wound dressings
- disposable gloves
- an advice leaflet.

All dressings, plasters, etc. should be individually wrapped. Dressings have a 'shelf life' and dates should be checked and anything out of date should be replaced because it may no longer be sterile.

Is an employer responsible for providing first aid facilities for members of the public, customers etc.?

No, the law on first aid applies to employees while they are at work.

However, it is good practice to consider the needs of customers or the public when completing the risk assessment. The Health and

Safety Executive strongly recommend that public and customers are included in first aid provision.

Are there any other responsibilities which an employer has in respect of first aid?

An employer must inform their employees of their first aid arrangements. This is a legal requirement.

Notices can be displayed advising where the first aid kit is located, who the appointed persons or first aiders are.

Special arrangements will need to be considered for any employees with language problems, learning disabilities, etc.

When is it necessary to provide a first aid room?

There is no legal requirement within the Health and Safety (First Aid) Regulations 1981 for employers to provide a first aid room. As with the provision of any first aid facilities, the need for a first aid room will be determined in the risk assessment.

Guidance in the Approved Code of Practice indicates that it would be good practice to provide a first aid room when there are 150 employees or more.

If a first aid room is provided, it must have the following attributes:

- be easily accessible to all employees
- be easily accessible for emergency services and for stretchers, etc.
- be provided with heating, lighting and ventilation
- be provided with hand-washing facilities and, preferably, a sink
- be provided with drinking water
- be provided with a chair, couch, table/desk, etc.
- be provided with first aid materials — first aid kits, etc.
- have a refuse bin

- contain blankets and pillows, etc.
- have some method of raising the alarm
- contain record books
- have surfaces which are easily cleaned and disinfected
- be kept clean and tidy
- be provided with suitable first aid information, e.g. names of first aiders, etc.

A first aid room should be exclusively used as a first aid room so that it is available in any emergency but if it needs to be used as a workplace, procedures should be in place to ensure that its use as a first aid room will not be prejudiced.

What records in respect of first aid treatment, etc. need to be kept by the employer?

It is considered good practice to keep records of all incidents which require any first aid treatment or attendance by a first aider.

A record book should be available to record the following:

- date, time and place of the incident
- injured person's name and job title
- a description of the injured person's injuries or illness
- details of first aid treatment given
- details of any actions taken after treatment given, e.g. did employee or person go to hospital, go home, etc.
- name and signature of person who gave first aid treatment or who oversaw the incident.

Consideration must be given to any data protection requirements and to ensure the anonymity of personal information. It would be a sensible idea to have a new page for each person treated and for previous entries to be kept in a secure drawer, etc. People should not be able to 'flip through' the first aid record book and learn personal facts about colleagues or others.

Case study

A young employee arrived at work in a call centre and, after about an hour or so, she reported that she did not feel well, had a headache and felt sick. The supervisor called the first aider who suggested that she would probably benefit from a 'lay down' in the first aid room. She was taken to the first aid room and settled down on the couch, still feeling quite poorly.Her colleagues were quite busy and forgot all about her being in the rest room until about lunch time. A colleague went to visit the first aid room and found that she was in a coma. An ambulance was called and the young woman was taken to hospital. Unfortunately, she died later that day from a brain haemorrhage.

Could anything have been done differently?

Yes, the first aider who attended the young woman should have been responsible for her care while she was in the first aid room and should have visited her every 30 minutes or so. If she had not felt better after, say, an hour, she should have been taken to her GP or to the casualty department at hospital. Records of the checks carried out on her should have been kept. She might still have died, but she should not have been left for several hours unattended.

13

Work-related upper limb disorders

What is a work-related upper limb disorder?

Work-related upper limb disorders (WRULD) are conditions which can affect the:

- neck
- shoulders
- arms
- elbows
- wrists
- hands
- fingers.

The symptoms include:

- aches and pains
- difficulty in movement
- swelling
- stiffness of joints.

Upper limb disorders can be recognised as industrial injuries and include:

- carpel tunnel syndrome
- tendonitis

- bursitis
- tripper finger
- vibration white finger.

Are work-related upper limb disorders the same as repetitive strain injuries?

Yes, generally anything which causes symptoms to the upper limbs and is caused by repeatedly undertaking the same tasks in the same way will fall into the category of upper limb disorders or repetitive strain injuries.

A wider term for such injuries is 'musculo-skeletal disorders'.

What parts of the body are affected?

Usually, the neck, shoulders, arms, wrists, hands and fingers. In particular, the muscles, tendons, ligaments, nerves or soft tissue associated with these joints.

What causes work-related upper limb disorders?

Upper limb disorders can occur in jobs which require repetitive finger, hand or arm movements, twisting movements, squeezing, hammering or pounding, pushing, pulling, lifting or reaching movements.

Both office-based jobs and manual activity can lead to WRULDS, e.g.:

- repetitive assembly line work
- inspection and packing
- meat and poultry preparation
- keyboard work

- supermarket checkout operation
- construction work.

The risk of WRULD is higher when there is:

- repetitive work processes
- force
- awkward posture
- insufficient rest breaks to allow muscles, etc. to recover.

As an employer, do I need to complete a risk assessment to establish whether employees are exposed to WRULD?

Yes. Any job task or activity where there is a risk to an employee's (or others') health and safety must be subject to a risk assessment.

Upper limb disorders are now quite well known and employers should be aware that they affect employees who carry out repetitive jobs of all types.

The general risk assessment undertaken for all work activities as required by the Management of Health and Safety at Work Regulations 1999 will suffice for identifying WRULDs.

There is no specific legislation which deals with work-related upper limb disorders, other than the Health and Safety (Display Screen Equipment) Regulations 1992 which require specific risk assessments to be carried out for work with display screens and their associated equipment.

How should work-related upper limb disorders be tackled?

The HSE have published a guidancc booklet on Upper Limb Disorders (HSG 60) and advocate a seven stage management approach as follows:

Step 1. Understand the uses and commit to action.
Step 2: Create the right organisational environment.
Step 3: Assess the risks of ULDs in the workplace.
Step 4: Reduce the risks of ULDs.
Step 5: Educate and inform the workforce.
Step 6: Manage any incidents of ULDs.
Step 7: Carry out regular audits and monitor the effectiveness of the programme.

Step 1

Everybody should understand that job tasks or activities could cause upper limb disorders, know how to identify the symptoms, how to report occurrences, how to adapt working processes, etc.

Management need to recognise that WRULDs are a significant hazard in the workplace and that they have a responsibility to reduce or eliminate them. To have employees off sick with WRULDs can be expensive, as can temporary replacements, inefficient working due to lower performance, etc.

Dealing with WRULDs is not necessarily expensive.

Operate a 'zero tolerance' policy for WRULDs.

Step 2

Create an environment where management are seen to recognise the hazards, risks and costs of poor working practices which lead to WRULDs.

Involve the workforce in seeking their own solutions.

Produce policies and procedures on how to undertake jobs, when to have breaks, acceptable behaviour, etc.

Ensure that employees who complain that they are suffering symptoms are not ridiculed. Listen to concerns and create a supportive atmosphere showing that you are prepared to act and resolve issues.

Step 3

Carry out regular and effective risk assessments on job tasks. Make sure that the persons carrying out the risk assessments are competent.

Step 4

Reduce the risk of WRULDs.

A process of risk reduction should be undertaken using a number of approaches, e.g. ergonomics.

Endeavour to eliminate the risk of injuries — change working practices, tools, posture, positions, etc.

Involve work groups in seeking solutions — often worker participation brings a positive 'buy in' attitude to solutions and they become more effective.

Review accident and sickness records. Are the same or more people off sick with limb disorders, pain, etc. Do people have only 1 or 2 days off and use 'pain in my wrist' or similar as the reason for absence.

Step 5

Provide information, instruction and training to employees.

What are the hazards of repetitive tasks? What are the symptoms? How can job processes, etc. be improved?

Ensure that employees know about the risk assessments and the physical control measures proposed.

Step 6

Respond to any complaints of limb injuries or disorders. Review the work in hand for the individual. Change things immediately.

Work-related upper limb disorders need not be permanent and can be reduced or eliminated if early action is taken. Continued strain on already strained muscles and tendons, for example, will make the injury worse, whereas stopping the task as soon as the pain is identified will enable the muscles or tendons to repair themselves and heal.

Step 7

As with any health and safety management programme, the control measures introduced need to be audited and reviewed to ensure their effectiveness.

Regular monitoring checks should take place to check that employees are following the correct procedures, etc.

Any deviations, new tools, adapted tools, etc. need to be accessed to establish why they are needed and how they can be used safely.

What are the steps for an effective risk assessment for work-related limb disorders?

Risk assessment for WRULDs is best carried out in two stages:

(1) identifying problem tasks
(2) assessing the risks.

It is essential to identify all problem, or likely problem, tasks in the organisation.

Brainstorm the tasks that everyone does. Walk the shop floor, office, department store, etc. and watch what people do.

Are people carrying out repetitive tasks quickly such as at supermarket checkouts or are the tasks less frequent but involving heavier weights or more awkward postures.

Record the basic features of the task on a checklist.

Review sickness and absence records.

Talk to employees — do they have any pain in their limbs, have general aches, etc.

How long do people carry out tasks without stopping.

WRULD: INITIAL ASSESSMENT

Company:		
Department:		
Initial assessment completed by:		
Date:		
Job task being assessed:		
Describe work activity/process:		
	Yes	**No**
Does the job involve:		
Gripping tools?		
Squeezing tools or implements?		
Twisting?		
Reaching?		
Moving things?		

	Yes	No
Lifting things?		
Finger/hand movement?		
Fast, short movements?		
Awkward posture?		
The use of force?		
Repetition of task?		
Does the employee:		
Have control of their work flow?		
Have to work to the speed of others?		
Take regular breaks?		
Are there:		
Actual cases of WRULDs?		
Complaints from employees about pains in hands, wrists, arms, etc.?		
Homemade, improvised changes to equipment, workstations or tools?		
Any other comments:		

Is a FULL RISK ASSESSMENT REQUIRED? Yes/No

(This will be YES if any of the above questions/statements were answered 'yes').

WRULD: FULL RISK ASSESSMENT

Name of company:	
Name of assessor:	
Date of assessment:	
Name of employee being assessed:	
Job title:	
Job task or activity:	
Describe task in detail:	
Repetition	
Does task involve repetition?	Yes/No
Describe, including frequency of movement, etc.	

Is repetition continuous for the shift or interspersed with breaks?	Yes/No
Is the job task always the same?	Yes/No
What could be done to reduce repetition?	

Working posture

Describe the working posture:

	Yes	**No**
Is the wrist bent in any way?		
Does the posture look awkward?		
Is a grip needed?		
Does the hand easily span the distance?		
Are tools awkward to hold or handle?		
Are tools turned or twisted?		
Are tools or equipment too high for the worker?		
Are tools or equipment too low for the worker?		
Can the worker reach things?		
Is there a lot of overhead reaching?		

	Yes	No
Are weights held with outstretched arms?		
Does the head and neck have to be put in awkward positions?		
Is there any visual sign of discomfort?		
What could be done to improve the working posture?		
Force		
Does the work activity require force of any kind, including pushing, pinching, twisting, hammering, need holding against the body, etc.? If yes, describe any type of force noted:		
What can be done to reduce the need for force?		
Working environment		
Describe the working environment, heating, lighting, ventilation, overcrowding, obstructions, noise, confined spaces, cramped working space, etc.		

Does anything add to the discomfort of the employee or create the need for more repetition of job tasks?		
What can be done to improve the work environment?		
Psycho-social factors		
What factors may affect the wellbeing of employees and what could contribute to poor working practices, repetition of tasks, etc.?		
	Yes	**No**
Do employees have control of their work pattern?		
Do they have to work to the speed of others?		
Can they have breaks?		
Are they on piece work?		
What can be done to improve conditions?		

What other factors could affect the health, safety and wellbeing of the employee or the person carrying out this task?		
ACTION PLAN		
What needs to be done?	By whom	When
Risk assessment reviews:		
Completion of action plan:		
General review:		
Signed:		
Position:		
Date:		

What are some of the control measures which may help to reduce WRULDs?

Control measures to reduce or eliminate work-related upper limb disorders can be extremely wide ranging and really are best judged for each job task or activity individually.

Any improvement in reducing repetitive jobs, awkward postures, carrying of weights, etc. will help to reduce risks of injury.

Some control measures which might be considered are:

- to mechanise or automate the process
- modify the operation or process
- change the shape of tools
- change the height of chairs, or their style
- reduce packaging size
- reduce weights
- reduce reach distances
- lower shelving
- provide mechanical aids
- reduce 'double handling'
- move tasks nearer to stock
- improve the environment
- create more work space
- give employees some control over their jobs
- introduce multi-tasking
- rotate employees
- adapt the job to the worker, not the worker to the job
- provide soft grip tools
- review PPE
- use lightweight tools
- improve lighting and other working environment factors.

14

Fire safety

What is the legal requirement for a fire risk assessment?

The Fire Precautions (Workplace) Regulations 1997, as amended in 1999, set down the requirements for employers to carry out a fire risk assessment for all workplaces.

Any premises in which persons are employed must be subject to a fire risk assessment.

The principles of risk assessment to be followed are those listed in the Management of Health and Safety at Work Regulations 1999.

Where there are five or more employees, the significant findings of the risk assessment must be recorded in writing.

Are premises with a Fire Certificate exempt from the need to carry out a fire risk assessment?

No. Premises with a Fire Certificate were originally exempt from completing fire risk assessments under the Fire Precautions (Workplace) Regulations 1997 but the European Commission declared that the UK had not implemented the European Directive correctly and required amendments to be made.

Amendments to the original 1997 Regulations were made in 1999 and a revised legal requirement was implemented which covered all workplaces. The 1997 Regulations are now read with the amendment Regulations.

The Management of Health and Safety at Work Regulations were also amended in 1999 and the requirement to complete a fire risk assessment was inserted into Regulation 3 on risk assessments.

What actually is a fire risk assessment?

A fire risk assessment is, in effect, an audit of your workplace and work activities in order to establish how likely a fire is to start, where it would be, how severe it might be, who it would affect and how people would get out of the building in an emergency.

A fire risk assessment should be concerned with *life safety* and not with matters which are really fire engineering matters.

A fire risk assessment is a structured way of looking at the hazards and risks associated with fire and the products of fire, e.g. smoke.

Like all risk assessments, a fire risk assessment follows *five key steps*, namely:

Step 1: Identify the hazards
Step 2: Identify the people and the location of people at significant risk from a fire
Step 3: Evaluate the risks
Step 4: Record findings and actions taken
Step 5: Keep assessment under review.

So, a fire risk assessment is a record that shows you have assessed the likelihood of a fire occurring in your workplace, identified who could be harmed and how and decided on what steps you need to take to reduce the likelihood of a fire (and therefore its harmful consequences) occurring. You have recorded all these findings regarding your undertaking into a particular format, called a risk assessment.

Definitions

Risk assessment — the overall process of estimating the magnitude of risk and deciding whether or not the risk is tolerable or acceptable.

Risk — the combination of the likelihood and consequence of a specified hazardous event occurring.

Hazard — a source or a situation with a potential to harm in terms of human injury or ill-health, damage to property, damage to the environment or a combination of these factors.

Hazard identification — the process of recognising that a hazard exists and defining its characteristics.

What are the *five steps* to a fire risk assessment?

Step 1. Identify the hazards

Sources of ignition

You can identify the sources of ignition in your premises by looking for possible sources of heat that could get hot enough to ignite the material in the vicinity.

Such sources of heat/ignition could be:

- smokers' materials
- naked flames, e.g. candles, fires, blow lamps, etc.
- electrical, gas or oil-fired heaters
- Hot Work processes, e.g. welding or gas cutting
- cooking, especially frying

- faulty or misused electrical appliances including plugs and extension leads
- lighting equipment, especially halogen lamps
- hot surfaces and obstructions of ventilation grills, e.g. radiators
- poorly maintained equipment that causes friction or sparks
- static electricity
- arson.

Look out for evidence that any items have suffered scorching or overheating, e.g. blackened plugs and sockets, burn marks, cigarette burns, scorch marks, etc.

Check each area of the premises systematically:

- customer areas, public areas and reception
- work areas and offices
- staff kitchen and staff rooms
- store rooms and cleaners' stores
- plant rooms and motor rooms
- refuse areas
- external areas.

Sources of fuel

Generally, anything that burns is fuel for a fire. Fuel can also be invisible in the form of vapours, fumes, etc. given off from other less flammable materials.

Look for anything in the premises that is in sufficient quantity to burn reasonably easily, or to cause a fire to spread to more easily combustible fuels.

Fuels to look out for are:

- wood, paper or cardboard
- flammable chemicals, e.g. cleaning materials
- flammable liquids, e.g. cleaning substances, liquid petroleum gas
- flammable liquids and solvents, e.g. white spirit, petrol, methylated spirit

- paints, varnishes, thinners, etc.
- furniture, fixtures and fittings
- textiles
- ceiling tiles and polystyrene products
- waste materials and general rubbish
- gases.

Consider also the construction of the premises — are there any materials used which would burn more easily than other types. Hardboard, chipboard and blockboard burn more easily than plasterboard.

Identifying sources of oxygen

Oxygen is all round us in the air that we breathe. Sometimes, other sources of oxygen are present that accelerate the speed at which a fire ignites, e.g. oxygen cylinders for welding.

The more turbulent the air, the more likely the spread of fire will be, e.g. opening doors brings a 'whoosh' of air into a room and the fire is fanned and intensifies. Mechanical ventilation also moves air around in greater volumes and more quickly.

Do not forget that while ventilation systems move oxygen around at greater volumes, they will also transport smoke and toxic fumes around the building.

Step 2. Identify who could be harmed

You need to identify who will be at risk from a fire and where they will be when a fire starts. The law requires you to ensure the safety of your staff and others, e.g. customers. Would anyone be affected by a fire in an area that is isolated? Could everyone respond to an alarm, or evacuate?

Will you have people with disabilities in the premises, e.g. wheelchair users, visually or hearing impaired? Will they be at any greater risk of being harmed by a fire than other people?

Will contractors working in plant rooms, on the roof, etc. be adversely affected by a fire? Could they be trapped or fail to hear alarms?

Who might be affected by smoke travelling through the building? Smoke often contains toxic fumes.

Step 3. Evaluate the risks arising from the hazards

What will happen if there is a fire? Does it matter whether it is a minor or major fire? Remember that small fires can grow rapidly into infernos.

A fire is often likely to start because:

- people are careless with cigarettes and matches
- people purposely set light to things
- cooking canopies catch fire due to grease build-up
- people put combustible material near flames/ignition sources
- equipment is faulty because it is not maintained
- electrical sockets are overloaded.

Will people die in a fire from:

- flames
- heat
- smoke
- toxic fumes?

Will people be trapped in the building?

Will people know that there is a fire and will they be able to get out?

Step 3 of the risk assessment is about looking at what *control measures* you have in place to help control the risk or reduce the risk of harm from a fire.

Remember — fire safety is about *life safety*. Get people out fast and protect their lives. Property is always replaceable.

You will need to record on your fire risk assessment the fire precautions you have in place, i.e.:

- What emergency exits do you have and are they adequate and in the correct place?
- Are they easily identified and unobstructed?
- Is there fire-fighting equipment?
- How is the fire alarm raised?
- Where do people go when they leave the building — an assembly point?
- Are the signs for fire safety adequate?
- Who will check the building and take charge of an incident, i.e. do you have a Fire Warden appointed?
- Are fire doors kept closed?
- Are ignition sources controlled and fuel sources managed?
- Do you have procedures to manage contractors? (Remember that Windsor Castle went up in flames because a contractor used a blow torch near the curtains!)

Taking all your fire safety precautions for the premises into consideration, is there anything more that you need to do?

Are staff trained in what to do in an emergency? Can they use fire extinguishers? Do you have fire drills? Is equipment serviced and checked, e.g. emergency lights, fire alarm bells, etc.

Step 4. Record findings and action taken

Complete a fire risk assessment form and keep it safe.

Make sure that you share the information with staff.

If contractors come to site, make sure that you discuss *their* fire safety plans with them and that you tell them what your fire precaution procedures are.

Step 5. Keep assessment under review

A fire risk assessment needs to be reviewed regularly — about every six months or so and whenever something has changed, including layout, new employees, new procedures, new legislation, increased stock, etc.

Is there any guidance on assessing the risk rating of premises in respect of fire safety?

When completing fire risk assessments it is sensible to categorise *residual risk* for the premises into a risk rating category — normally referred to as high, medium or low.

In terms of fire risk rating it is usual to refer to medium risk as 'normal'.

The Government's publication *Fire safety — an employer's guide* gives some guidance on how to fire risk rate premises.

High risk premises

- Any premises where highly flammable or explosive substances are stored or used (other than in very small quantities).
- Any premises where the structural elements present are unsatisfactory in respect of fire safety:
  - lack of fire-resisting separation
  - vertical or horizontal openings through which fire, heat and smoke can spread
  - long and complex escape routes created by extensive sub-division of floors by partitions, etc.
  - complex escape routes created by the positioning of shop unit displays, machinery, etc.
  - large areas of smoke- or flame-producing furnishings and surface materials especially on walls and ceilings.
- Permanent or temporary work activities which have the potential for fires to start and spread, e.g.:
  - workshops using highly flammable materials and substances

    - paint spraying
    - activities using naked flames, e.g. blow torches and welding
    - large kitchens in work canteens and restaurants
    - refuse chambers and waste disposal areas
    - areas containing foam or foam plastic upholstery and furniture.
- Where there is significant risk to life in case of fire:
    - sleeping accommodation provided for staff, guests, visitors, etc. in significant numbers
    - treatment or care where occupants have to rely on others to help them
    - high proportions of elderly or infirm
    - large numbers of people with disabilities
    - people working in remote areas, e.g. plant rooms, roof areas, etc.
    - large numbers of people resorting to the premises relative to its size, e.g. sales at retail shops
    - large numbers of people resorting to the premises where the number of people available to assist is limited, e.g. entertainment events, banquets, etc.

Normal risk premises

- Where an outbreak of fire is likely to remain contained to localised areas or is likely to spread only slowly, allowing people to escape to a place of safety.
- Where the number of people in the premises is small and they are likely to escape via well-defined means of escape to a place of safety without assistance.
- Where the premises have an automatic warning system or an effective automatic fire-fighting, fire-extinguishing or fire-suppression system which may reduce the risk categorisation from high.

Low risk premises

Where there is minimal risk to peoples' lives and where the risk of fire occurring is low or the potential for fire, heat or smoke spreading is negligible.

What type of people do I need to worry about when I carry out my risk assessment?

Employers must consider the following people as being at risk in the event of a fire:

- employees
- employees whose mobility, sight or hearing might be impaired
- employees with learning difficulties or mental illness
- other persons in the premises if the premises are multi-occupied
- anyone occupying remote areas of the premises
- visitors and members of the public
- anyone who may sleep on the premises.

Does a fire risk assessment have to consider members of the public?

A fire risk assessment must be carried out by an employer and must consider the risks to the safety of *employees*.

However, under the general provisions of the Management of Health and Safety at Work Regulations 1999, all persons who may be affected by the employer's business or undertaking must be considered in the risk assessment.

Case study

Fire hazards in licensed premises

Fires in public houses are quite common and usually occur in the kitchen or customer area where the risks of fire are greatest.

What do I need to look for?

Kitchen

- Storage of flammable materials, e.g. cardboard packaging near to a heat source.
- Grease build-up on grills, filters, cooking equipment as it can easily ignite and flames can spread rapidly.
- Grease and dirt build-up on canopies and within ductwork.
- Overheating deep fat fryers — oil left burning for long periods of time.
- Flammable materials next to cooker, grill, griddle or hob tops.
- Overloaded sockets.
- Poorly repaired plugs.
- Use of wrong fuses in plugs.
- Poor use of extension leads, wrong fuses or overloaded sockets.
- Poorly maintained equipment.
- Misuse of microwaves and use of combustible material as packaging, cardboard dishes, etc.
- Storage of flammable aerosols and cleaning fluids near to heat sources.
- Use of blow torches for 'burnishing' toppings, glazing, etc.

- Generation of flammable fumes and vapours from aerosols, etc. used elsewhere but where the fumes drift and ignite near a heat source.
- Use of LPG in the kitchen.
- Faulty gas appliances — usually associated with explosions.
- Arson.

Bar servery

- Overloaded sockets.
- Disposal of cigarette debris, ashtrays, etc. into rubbish bins or waste bins which are combustible and which contain combustible material, e.g. packaging.
- Electric faults on equipment, e.g. glass washers and fridges.
- Spread of fire to the bar via python runs and other voids which communicate with the other areas of the pub.
- Overheating equipment due to vent grill obstructions.

Customer area

- Smoking materials dropped onto seating areas, carpeting, etc.
- Poorly extinguished cigarettes.
- Disposal of ashtrays into waste receptacles containing combustible materials.
- Overloaded electrical sockets.
- Furniture, etc. too close to real fires or gas fires.
- Light fittings with the wrong wattage bulbs, etc.
- Light fittings too near to combustible objects where heat transference can cause combustion.
- Arson.

Cellar

- Overloaded electrical sockets.
- Use of flammable cleaning fluids near electrical ignition sources.
- Storage of flammable substances near heat sources.
- Discarded smoking materials, i.e. hastily discarded cigarettes, matches, etc.
- Storage of combustible materials near a heat source.
- Blocking up equipment ventilation grills, e.g. ice machines, causing the equipment to overheat and spontaneously combust.
- Arson.

Plant rooms

- Poorly maintained equipment.
- Storage of combustible materials and substances near to heat sources.
- Electrical faults.
- Escape of combustible fumes, gases, etc.
- Overloaded electrical sockets and extension leads.
- Electrical arcing.
- Use of flammable cleaning chemicals.
- Grease and dirt build-up within equipment which is heat generating.

Staff areas

- Many of the incidents already listed can apply to staff areas.
- Faulty washing machine and tumble drier equipment.
- Poor ventilation to electrical equipment.
- Use of portable heating equipment.

Top tips

- Make regular checks to identify fire hazards.
- Look out for anything unusual — blackening of plugs, sockets, etc.
- Do *not* overload electrical sockets.
- Undertake regular maintenance of equipment.
- Be 'Fire Aware'.
- Train staff to be 'Fire Aware'.

Who can carry out a fire risk assessment?

The Fire Precautions (Workplace) Regulations 1997 (amended) state that the person who carries out a fire risk assessment shall be *competent* to do so.

Competency is not defined specifically in the Regulations, or in the Management of Health and Safety at Work Regulations 1999.

Competency means having a level of knowledge and experience which is relevant to the task in hand.

Many fire authorities, who enforce the Fire Precautions (Workplace) Regulations 1997 (amended) do not advocate that consultants are employed to carry out complicated assessments.

A fire risk assessment is a logical, practical review of the likelihood of a fire starting in the premises and the consequences of such a fire. Someone who has good knowledge of the work activities and the layout of the building, together with some knowledge of what causes a fire to happen, would be best placed to carry out a fire risk assessment.

What sort of fire hazards need to be considered?

Consider any significant fire hazards in the room or area under review:

- combustible materials, e.g. large quantities of paper, combustible fabrics or plastics
- flammable substances, e.g. paints, thinners, chemicals, flammable gases, aerosol cans
- ignition sources, e.g. naked flames, sparks, portable heaters, smoking materials, Hot Works equipment.

Do not forget to consider materials which might smoulder and produce quantities of smoke. Also, consider anything which might be able to give off toxic fumes.

Consider also the type of insulation involved or used in cavities, roof voids, etc. Combustible material may not always be visible, e.g. hidden cables in wall cavities.

What sort of structural features are important to consider when carrying out a fire risk assessment?

Fire, smoke, heat and fumes can travel rapidly through a building if it is not restricted by fire protection and compartmentation.

Any part of a building which has open areas, open staircases, etc. will be more vulnerable to the risk of fire should one start.

Openings in walls, large voids above ceilings and below floors allow a fire to spread rapidly. Large voids also usually contribute extra ventilation, thereby adding more oxygen to the fire.

A method of fire prevention is to use fire-resistant materials and to design buildings so that fire will not travel from one area to another.

Any opportunity for a fire to spread through the building must be noted on the fire risk assessment.

What are some of the factors to consider when assessing existing control measures for managing fire safety?

Many premises and employers already have some level of fire safety management in place and the Fire Precautions (Workplace) Regulations 1997 (amended) were not intended to add an especially heavy burden onto employers.

Existing control measures must be reviewed and the following are examples of what to look for:

- the likely spread of fire
- the likelihood of fire starting

- the number of occupiers of the area
- the use and activity undertaken
- the time available for escape
- the means of escape
- the clarity of the escape plan
- effectiveness of signage
- how the fire alarm is raised
- can the alarm be heard by everyone
- travel distances to exits
- number and widths of exits
- condition of corridors
- storage and obstructions
- inner rooms and dead ends
- type of staircases and access to staircases
- openings, voids, etc. within the building
- type of fire doors
- use of panic bolts
- unobstructed fire doors
- intumescent strips
- well fitting fire doors
- propped open fire doors
- type of fire alarm
- location, number and condition of fire extinguishers
- display of fire safety notices
- emergency lighting
- maintenance and testing of fire alarm break glass points
- installation of sprinklers
- location and condition of smoke detectors
- use of heat detectors
- adequate lighting in an evacuation
- training of employees
- practised fire drills
- general housekeeping
- management of contractors
- use of Hot Works Permits
- control of smoking

- fire safety checks
- provision for managing the safety of people with disabilities
- special conditions, e.g. storage of flammable substances
- storage of combustible materials near a heat source.

The best fire risk assessments are 'site specific' — review and inspect your *own* workplace.

Checklist

What should a fire risk assessment cover?

- Identification of hazards.
- Sources of ignition.
- Identification of persons at risk from fire.
- Means of escape from the building.
- Fire warning systems.
- Fire-fighting facilities.
- Identification of fire safety procedures, i.e. emergency procedures.
- Review of the controls in place and recommendations for improvements where necessary.

The following pages contain a number of different types of fire risk assessment formats.

FIRE RISK ASSESSMENT

Name of premises: ______________________________

What particular area are you reviewing for this Fire Risk Assessment?

What activity, practice, operation, etc. are you particularly reviewing for this Fire Risk Assessment?

What ignition sources have you identified?

What sources of fuel have you identified?

Are there any 'extra' sources of oxygen, or will mechanical ventilation increase oxygen levels?

Does anyone do anything that will increase the risk of a fire starting?

If a fire were to start, who would be at risk?

Would anyone be at any extra or special risk, or would any injuries/ consequences of the fire be increased?

What precautions are currently in place to reduce the likelihood of a fire occurring, or to deal with it/control it if a fire did start (e.g. checks, alarms, emergency procedures, etc.)?

What other precautions need to be taken, if any? Does anything need to be done to improve existing fire precautions?

How will the information in this Risk Assessment be communicated to staff? Will any staff training take place?

Is there anything else that you think needs to be recorded on this Risk Assessment?

After having identified the hazards and risks of a fire starting and after considering all the procedures you *currently* have in place, do you consider the risk to *life safety* of either staff or customers (including any contractors, visitors, etc.) to be:

High ☐ Medium/normal ☐ Low ☐

If risks to life safety are very likely or possible, steps MUST be taken to improve fire safety.

If you implement the other, additional measures identified in this Fire Risk Assessment, will risk to *life safety* of either staff or customers (plus others) be:

High ☐ Medium/normal ☐ Low ☐

If risks to life safety are possible or very likely, then greater control measures MUST be implemented.

Risk Assessment completed by: ______________________________

Date: ______________________________

Fire Risk Assessment needs a review on: ______________________________

FIRE RISK ASSESSMENT

Name and address of premises: ______________________________

____

____

Owner/Employer/Person in Control: ______________________________

Contact details: ______________________________

Date of Risk Assessment: ______________________________

Completed by: ______________________________

Use of premises/area under review: ______________________________

Identification of fire hazards	High	Medium	Low

Identification and location of those at risk
Evaluation of the risks
Significant findings
Actions taken to reduce/remove risks
Residual Risk Assessment High ☐ Medium/normal ☐ Low ☐
Review of Risk Assessment: Under what circumstances: How often:

FIRE RISK ASSESSMENT

Name and address of premises:				
Area/room/floor under assessment:				
Name of person completing assessment:			Date of assessment:	
Fire hazard	**People at risk**	**Existing control measures**	**Proposed action to be taken**	**Date action taken and by whom (include signature)**

EXAMPLE OF COMPLETED FIRE RISK ASSESSMENT

Name of premises: Anywhere Hotel			
Date of Assessment: June 2002			
Name of person carrying out Assessment: Perry Scott Nash Associates Limited			
Area of premises being Assessed: Restaurant, Kitchen, Bar, Champagne Bar, Production Kitchen, Pastry Kitchen and Bedrooms.			
Identification of Fire Safety Hazards	Yes	No	Don't know
Combustible materials, furnishings, e.g. carpets, curtains, seat coverings	✓		
Smokers and discarded smoking materials	✓		
Electric appliances, e.g. portable equipment	✓		
Storage of chemicals, flammable aerosols, etc.	✓		
Portable heaters, LPG, flambé lamps	✓		
Cooking equipment, ducts, gas pipes	✓		
Accumulations of waste, including cooking oil	✓		
Obstruction of vents or cooling systems		✕	
Building works		✕	
Items stored too near to heat sources		✕	
Obstruction of fire exit routes or fire doors propped open		✕	

Comments (Please describe anything highlighted 'yes' above)			
Furnishings and fittings throughout the premises, including public areas, bars, restaurants, event rooms, bedrooms and offices. Smokers are permitted within the premises and smoking waste is controlled by discarding smoking materials into designated ashbins. Electrical appliances including televisions, kettles, irons, hairdryers, glass washers, microwaves, other kitchen appliances and computers. Chemical storage for cleaning products and storage of CO_2. Flambé lamps used within the restaurants. Cooking equipment includes gas oven stoves, grills and deep fat fryers in all kitchens and canopy extract systems. Accumulation of waste in bar, kitchen, housekeeping and refuse area including discarded smoking material, paper waste and collection of waste cooking oil.			
Who might be at risk from the above hazards: customers: • to pub • to hotel staff disabled people contractors?	 ✓ ✓ ✓ ✓ ✓ ✓		

Are any people *particularly* at risk if a fire should break out, e.g. people working alone in plant rooms, visitors, etc.? Describe how and why they will be at risk: Controls are in place to ensure contractors and visitors are accounted for. Procedure in place indicates person's name, time of arrival, area of work, type of work, expected time of work/visit, and each person signs out before leaving. However, some contractors might leave the hotel without informing security. Disabled/special needs persons accommodated on the first and second floor. In the event of a fire the lift cannot be used and only means of escape is via staircase. Special needs rooms — 161, 167, 205 and 207; all front desk staff are aware of occupants' needs.			

WHAT FIRE SAFETY CONTROL MEASURES ARE ALREADY IN PLACE

CONTROL MEASURE	DESCRIBE
Fire alarm — warning	Alarm bell activated either by fire call alarm points, automatic smoke and heat detection, Stratos — high sensitivity photo electric smoke detector OR by activating the main fire alarm panel. A continuous ringing throughout the premises, which can only be silenced at the alarm panel. All fire call alarm points are clearly signed throughout the premises.
Fire alarm — detection	Automatic heat and smoke detectors located throughout the premises. The Hotel is a grade two listed building and high sensitivity smoke detectors are situated within the fabric of the building.
Escape lighting	Emergency lighting luminaire units installed over all fire exit routes indicating escape routes and also installed within escape routes located from the ceiling using directional signage to define escape route.
Escape routes	From lower and upper basement via internal metal staircase to ground level. From ground, first, second, third, forth, fifth and sixth floor areas via internal routes including metal staircase or by external metal staircase. All routes are segregated by automatic fire doors, which close in the event of a fire to prevent fire spread.

Emergency signage	Directional signage clearly defines escape routes. Staff fire action route signage located throughout the premises with pictorial and written instruction.
Fire extinguishers	Fire extinguishers and fire hoses located on every level of the hotel. Fire extinguishers include water, foam, powder and CO_2 (refer to Appendix 1 for type of fire extinguisher and location) also (refer to Appendix 2 for fire hose location).
Fire blanket	Nine fire blankets located within all kitchens at the Hotel. (Refer to Appendix 3 for fire blanket location.)
Testing of the system — what records are kept?	Fire call alarm points are tested every Friday at 16:00. Monthly testing of fire extinguishers, fire exit/escape and emergency lighting. A competent contractor undertakes annual tests of fire extinguishers, last tested in January 2002, Ansul system last tested 30 April 2002, lighting is ongoing annually (refer to Appendix 4 for electrical testing). All monthly and annual tests are recorded and appropriate certification is kept.
Fire escape routes — how are they maintained clear of obstructions, how often are they checked, etc.?	Staff members are informed to keep all fire escape routes clear including corridors and fire exit doors. Daily and nightly checks are carried out to ensure routes are not obstructed. CCTV allows security to permanently check routes. Any obstructions are logged and removed.

How do people with disabilities hear the alarm, get out of the building? (Include hotel guests.)	Front desk staff keep details of guests with special needs. In the event of a fire, a disabled person would be assisted down via the staircase if safe to do so. Staff would assist and guide other guests/patrons to the assembly point, on route to leaving the premises.
Who calls the Fire Brigade?	Management and senior staff are responsible for formulating the emergency plan in the event of a fire. Front desk staff would call the Fire Brigade under instruction.
Security — are all fire exits/fire doors openable, kept unlocked during opening hours, etc.?	All fire exits/fire doors have push bar operation or turn knobs to fully open doors in the event of a fire. No fire door is locked or requires a key to open it in the event of a fire.
What procedures are put in place at night to ensure that fire hazards are managed and evacuations can be undertaken successfully?	Night staff undertake checks throughout the premises to include that fire doors are closed, corridors/fire routes are not obstructed and for signs of fire. Currently, no formal procedure implemented, however any concerns are noted on the night check sheet and, if necessary, a digital photo is taken for evidence. This information is then given to security during handover. All night staff receive three-monthly fire safety training to include evacuation.

Are emergency evacuation notices displayed?	Each room has an evacuation notice with plan and instructions for the event of a fire; this is displayed within a frame on the back of the bedroom door. Staff fire action route signage is located throughout the premises with pictorial and written instructions to include: i) operating alarm call poin ii) calling the Fire Brigade iii) fire fighting iv) evacuation v) assembly point.
Have all staff been trained in what to do in the event of a fire?	All new staff receive induction fire safety training which is carried out every Monday. All daytime staff receive refresher fire safety training every six months and all night staff receive refresher fire safety training every three months. Evacuation training is carried out every six months with a full evacuation of the hotel. All staff sign for training and records are kept by the HR Department.
Who 'looks after' guests, contractors, etc.?	Management and staff look after guests in the event of a fire. A full occupation list is printed off and front desk staff check guests at the assembly point. Security look after contractors and visitors, full formal procedure of checking in and out of contractors and visitors.
What housekeeping procedures are in place to keep rubbish down, store chemicals safely, etc.?	All chemicals are stored in locked cupboards with full COSHH information for each chemical. All waste is removed to designated refuse area with frequent collection to remove build-up of waste.

Is equipment regularly maintained to ensure that it works effectively, thus reducing the risk of electrical failure?	Yes. Portable appliance testing (PAT) carried out annually. Maintenance record all tests carried out. All fixed equipment is tested by a competent person every two to three years and recorded/certificated with the Maintenance Department. Staff remain vigilant and report any defects to heads of department for action.
FURTHER COMMENTS	

WHAT ADDITIONAL CONTROL MEASURES ARE NEEDED?

1. Provide signage to emergency stop buttons situated within the kitchen to indicate gas, electricity or both. (**7 days**)
2. Remove cardboard boxes and other items stored within the electricity cupboard situated in the Bar and ensure that regular checks are made to prevent overstocked items obstructing electrical panels. (**Immediate**)
3. Provide designated ashbin within the staff canteen for discarding of waste smoking materials. (**Immediate**)
4. Provide correct signage and fix fire extinguishers A49 and A50 within the electrical plant room situated on lower basement. (**Immediate**)
5. Replace directional fire exit sign situated within the laundry corridor. It is recommended that a plastic sign be fixed to the wall higher than the top of contract laundry trolleys to prevent damage to the sign. (**7 days**)
6. Replace fire blanket number one in the staff canteen kitchen as this was found to be greasy and dirty during the Fire Risk Assessment audit. Ensure that all fire blankets are checked for condition and replace when necessary. (**7 days**)
7. It is unclear if the fire hoses have recently been checked by a competent person. Check with Fire Protection who carried out the recent examination during January 2002 and obtain records. (**7 days**)

8. Ensure all fire doors leading onto escape routes are fitted with intumescent strips along the edges of the door. The majority of fire doors have been fitted with intumescent strips but it was noted that several kitchen fire doors had not — investigate and complete works. (**7 days**)
9. It is strongly recommended that a fire-proof safe be installed to store all fire safety records, to prevent damage or loss in the event of a fire. (**1 month**)
10. Provide emergency button within the steam room situated in the Train gym. It is understood that the works are in hand — awaiting installation of emergency button.
11. Implement formal documentation for night checks carried out indicating time, location and person carrying out checks. This may be paper- or computer-based documentation. It is recommended that the night fire safety checks are carried out half hourly between midnight and 07:00, to ensure that all areas of the premises are checked. (**7 days**)
12. The metal fire staircase situated at the fire exit, has been in-filled with wooden blocks to prevent a trip hazard. The in-fills should be metal to provide integrity to the fire escape route. Investigate and replace the in-fills with metal to match existing fabrication. (**1 month**)
13. Review Fire Risk Assessment yearly or when any changes are made to fire-fighting equipment/detection or internal procedures to ensure that it reflects current operational procedures within the Hotel. (**Ongoing**)

Risk Assessment review date:

Risk Assessment to be reviewed annually or whenever circumstances change that necessitate reassessment.

EXISITING RISK RATING (WITH CURRENT CONTROLS)

Low ✓ Medium____ High____

PROPOSED RISK RATING (WITH RECOMMENDED CONTROLS WITHIN THIS RISK ASSESSMENT)		
Low ✓	Medium____	High____

15

Design risk assessments (CDM)

The CDM Regulations place responsibilities on 'designers'. Who are 'designers'?

Under the CDM Regulations, 'designers' are all those who have some input into design issues in respect of a project. These include:

- architects and engineers contributing to, or having overall responsibility for, the design
- building services engineers designing details of fixed plant
- surveyors specifying articles or substances or drawing up specifications for remedial works
- contractors carrying out design work as part of a design and build project
- anyone with authority to specify or alter the specification of designs to be used for the structure, including the client
- temporary works engineers designing formwork and falsework
- interior designers, shopfitters and landscape architects.

The above includes architects, quantity surveyors, structural engineers, building services engineers, interior designers, project managers (if they can change or issue specifications), landscape architects/designers, temporary works engineers designing propping systems, etc.

The 'designer' must be carrying on a trade, business or other undertaking in which he prepares a design for a structure. The wording was amended to re-instate the intent of CDM which had been overturned by a court case.

What are the responsibilities of a designer?

Designers from all disciplines have a contribution to make in avoiding and reducing health and safety risks which are inherent in the construction process and subsequent work, e.g. maintenance.

The most important contribution a designer can make to improve health and safety will often be at the concept and feasibility stage where various options can be considered so as to avoid potential health and safety issues.

Designers must therefore give due regard to health and safety in their design work.

Designers must provide adequate information about health and safety risks of the design to those who need it, e.g. proposed roof access routes, use of fragile materials, etc.

Designers must co-operate with the planning supervisor and other designers on the project and ensure that information is freely available regarding health and safety issues and that they consider the implications of their designs with other aspects of the design, e.g. structural works in relation to building services, etc.

Designers must advise clients of their duties under CDM, as specified in Regulation 13(1). CDM requires designers to take reasonable steps to advise their clients of the existence of CDM, their duties within the Regulations, the existence of the Approved Code of Practice, good health and safety management and the benefit of making early appointments.

What are 'design risk assessments'?

'Design risk assessment' is the technical term given to the formal consideration of health and safety issues relevant to your design.

As a designer, you should understand that the term 'hierarchy of risk control' is a step-by-step process to eliminating, minimising or controlling health and safety hazards and risks.

Step 1: Eliminate the hazard and risk.
Step 2: Minimise the hazard and reduce the risk.
Step 3: Control the hazard and risk at source.
Step 4: Control the hazard and risk within the workplace.

In simple terms:

Step 1: Do not design something which could cause injury or ill health.
Step 2: If you have to continue with the design, ensure that you include safety features.
Step 3: Provide safety features, etc. at the site of the hazard.
Step 4: Ensure that everyone is issued with personal protective equipment.

Design risk assessments set out how you have considered the health and safety aspects of your design and record your decisions.

The first rule of health and safety is to eliminate hazard and risk. If this is not possible, do the next best thing by designing in safety features or specifying different material, etc.

Design risk assessments do not have to be detailed or complicated forms which cover everyday situations. They need to be specific to the project and should highlight *unusual* design considerations.

Designers should not only consider *how* the structure will be built and identify these hazards and risks, but must also consider how the building will be used in the future for maintenance and cleaning purposes as well as how it will be occupied.

Case study

Designers of a new concept retailing outlet sourced a unique mesh-type ceiling material which was to be installed in one single sheet to give a ripple effect to the ceiling.

Design considerations included the following.

- How is it delivered to site?
- Will it be difficult to handle?
- Will the metal mesh have jagged edges?
- How will it be fixed to the ceiling frame?
- How will lighting be fixed?
- How will sprinklers be fixed?
- Will it be a fragile surface?
- How could maintenance personnel walk above it to access fittings?
- How will it be cleaned?

All of the above questions, and many more, formed part of the design risk assessment. The planning supervisor was asked to comment on the information available and to offer any advice.

The new ceiling concept was installed and created the innovative design which the designers and client had hoped for.

The considerations you have made in respect of health and safety can be included on your drawings, provided the information is clear and unambiguous.

Each project could have a simple design risk assessment sheet attached to the project file, indicating how you made the key health and safety decisions. An example of a design risk assessment form is shown at the end of this chapter.

DESIGN RISK ASSESSMENT

Project address:

Designer:

Design company:

Description of project:

Design activity under assessment:

Hazards identified	**Construction**	**Maintenance/cleaning**

Design consideration to eliminate or reduce hazard:

Residual risk:

Control measures necessary:

Information to be relayed to planning supervisor for inclusion in:

(i) Pre-Tender Health and Safety Plan

(ii) Health and Safety File

Other relevant health and safety information:

Signed: ....

(Designer)

Date: ....

Alphabetical list of questions